# Water-in-Plants Bibliography

# volume 3 1977

References no. 2480 - 3686 / ACE-ZYA

Editors J. Pospísilová and J. Solárová

Dr. W. Junk bv Publishers The Hague/Boston/London 1979

Contributors

J. Solárová

J. Pospísilová

Z. Sesták

J. Catsky

I. Tichá

D. Hodánová

ISBN-13: 978-90-6193-903-0          e-ISBN-13: 978-94-009-9642-7
DOI: 10.1007/978-94-009-9642-7

# PREFACE

The third volume of Water-in-Plants Bibliography includes papers in all fields of plant water relations research which appeared during the year 1977 - from theoretical considerations about the state of water in cells and its membrane transport to drought resistance of plants or physiological significance of irrigation. In addition to papers devoted entirely to plant water relations, papers on other topics are included if they contain data on plant hydration level, water vapour efflux, rate of water uptake or water transport, etc., or if they contain valuable methodological information (measurement of selected microclimatic factors, soil moisture, etc.).

We have tried to cover fully the relevant papers which have appeared in the important scientific periodicals and books. Articles published in local journals, mimeographed booklets, abstracts of thesis and of symposia contributions, etc., were chosen mostly from reprints received directly from authors. The courtesy of those authors who have already supplied us with reprints and lists of their publications is highly appreciated. The manuscript is usually prepared in May and June of the year following the year which it covers. Unfortunately some reprints come later and thus the respective references appear in the following volume, with one year delay.

To maximize the value of the bibliography the references are arranged alphabetically according to the authors' names, and each volume is provided with three indexes. The authors' index contains all names of authors, co-authors and editors. Plant genera used as experimental material are indexed according to their Latin names. The subject index covers primary items chosen according to the interest of water relation researchers. Its preparation was based not only on the titles, key words and abstracts but also on the whole content of the article.

Since more than 1000 relevant papers dealing with plant water relations are currently published every year and included in this bibliography, and since all citations have been checked with the originals, collecting and preparing for publication such a large amount of material would have been impossible without collaboration of our colleagues from the Department of Physiology of Photosynthesis and Water Relations of the Institute of Experimental Botany of the Czechoslovak Academy of Sciences in Prague. We have also acknowledge with thanks the cooperation of Mrs. Drahomíra TĚŽKÁ Mrs. Vlasta FLORIÁNOVÁ, Mrs. Lenka KOLČABOVÁ, Mrs. Marie MANDLOVÁ, Mrs. Eva PLESSINGEROVÁ, Mrs. Marta ŠMÍDOVÁ and Miss Dana ŠIŠKOVÁ who helped in typing card material, preparing indexes, etc., and Mr. Petr ZÁZVORKA who supplied us with rare periodicals.

Dr. Jana Pospíšilová and Dr. Jarmila Solárová

Institute of Experimental Botany
Czechoslovak Academy of Sciences

Flemingovo nám. 2
160 00 PRAHA 6
Czechoslovakia

Praha, 7 November 1978.

## INSTRUCTIONS FOR USE

All references are arranged alphabetically according to the authors' names. They are numbered and these numbers are used in the indexes.

Authors' names are presented in the spelling used in the original paper. If this spelling does not correspond to the spelling usually used by the author (e.g. Russian papers of English authors), one spelling is referred to the other in the Authors' Index. Like the transcriptions they are alphabetically arranged mostly according to the authors' own references. Nevertheless, the editors apologize for some misinterpretations which are partly corrected by the cross-indexing in the Authors' Index.

The references contain the original unshortened title of the paper (book). English, French, and German titles are cited in the original language. Titles in other languages are supplemented with a translation in English (using the title of the respective English abstract, if it is presented). Titles of Japanese, Chinese etc. papers are given in English translation only. In both these cases the abbreviations of the original language and the language of the abstract are given in brackets at the end of the reference. The following abbreviations are used most frequently:

| | |
|---|---|
| Belorussian | Japanese |
| Bulgarian | Latvian |
| Chinese | Lithuanian |
| Croatian | Norwegian |
| Danish | Polish |
| English | Russian |
| Esthonian | Roumanian |
| French | Slovak |
| German | Spanish |
| Georgian | Swedish |
| Hungarian | Ukrainian |
| Italian | Uzbeg |

The transliteration of Cyrillic characters is in accordance with the BSI-ASA/SC-Z39 draft table, i.e.:

| | | | |
|---|---|---|---|
| a | а | p | п |
| b | б | r | р |
| ch | ч | s | с |
| d | д | sh | ш |
| e | е | shch | щ |
| e | э | t | т |
| f | ф | ts | ц |
| g | г | u | у |
| i | и | v | в |
| j | й | y | ы |
| k | к | ya | я |
| kh | х | yu | ю |
| l | л | z | з |
| m | м | zh | ж |
| n | н | " | ъ |
| o | о | ' | ь |

Several exceptions apply for Ukrainian and Belorussian.

| | | |
|---|---|---|
| Ukrainian: | y | и |
| | i | і |
| | ï | ї |
| Belorussian: | ŭ | ў |

The journals' names are abbreviated mainly according to the Style Manual for Biological Journals (Second Edition, Amer. Institute of Biological Sciences, Washington, D.C. 1964), e.g.:

Abhandlungen
Abstract
Abteilung
Academy
Acker
Acta
Advances
Africa (-ican)
agricultural
Agriculture
Agrobiologia (-ogy)
Agrobotanica
Agrokémia
Agronomy
agropecuaria
Akademie (-emiya)
Algology
allgemeine
America
American
Anais (-alele)
Analysis
analytical
Anatomy
angewandte
animal
Annales (-als)
annual
anorganic (-anisch)
applied
aquatic
Arbeit
Archiv
Argentina
Atmosphere
atmospheric
atomic
Australia (-ralian)
Azerbaĭdzhanskaya
Bacteriology
Beiheft
Beiträge
Belgique
Belorusskaya
Berichte
biochemical
Biochemie
Biochemistry
biochimica
biokhimicheskiĭ
Biokhimiya
Bioklimatologie
Biologia (-ogy)
biological (-ogisk)
biophysical
Biophysics
Bodenkunde
Boletin (-ettino)
Bolgarskiĭ
botanica (-anicorum)
botanical (-anisca)
Botanika (-any)
Brasileira
Brazil

Contamination
Contribution
Control
Conservation
Croatica
cultural
Culture
Current
Cytobiology
Cytochemistry
Cytology
Czechoslovak
Danske
dendrological
Dendrology
Department
Deutschland
Development
Disease
Dissertation
Division
Doklady
Dopovidi
Drainage
ecological
Ecology
Economy
Edafology
Education
Ékologiya
éksperimental'nyĭ
Embryology
Encyclopedia
Engineering
Enology
Entomology
environmental
Enzymology
Estonskaya
European
Experiment
experimental
Faculty
Federation
Fizika
Fiziologiya
Flurbereinigung
forestiere
Forestry
Forschung
Foundation
France
Gazette
general
genetical
Genetics
Genetika
Geobotany
Geofizika
Geophysics
Gesellschaft
Giornale
Governement
Grassland

Journal
Khimiya
Klasse
Kongelige
Közlemenyek
kul'turnykh
Laboratory
Landbauforschung
Landwirtschaft
lesní (-iho)
Letters
Limnology
Linnean
Litovskoĭ
Lucrarile
Magazin
Management
marina (-ine)
Material
Mathematics
Mededelingen
Mediterranean
Meldinger
Meteorology
Microbiology
Midland
Modeling
molecular
Monographiae (-aphy)
Moskovskiĭ
Mycology
national
natural
Naturalist
naturelle
naturkundliche
Naturforschung
nauchnye
Neerlandica
Netherland
New Zealand
Norges
Norwegian
Notiser
nuclear
Nutrition
obshcheĭ (-iĭ)
Oceanography
Oecologia
Ökologie
Optics
organic
original
ornamental
Otdelenie
Paleobotany
Palynology
Pathology
Pesquisa
Pesticide
Pflanzen-
Pflanzenernährung
Pflanzenphysiologie
Pflanzenzüchtung

Rastenievodstvo
Recherche (-erches)
Report
Research
Resources
Review (-ista, -ue)
Rivista
Roczniky
rolniczych
Rostlin (-lina)
rostlinná
Roumaine
royal
Russian
Russkiĭ
Sbornik
Scandinavica
Scandinavicus
School
Science
scientific
Section
Selektsiya
Selskabs
sel'skokhozyaistvennyĭ
Series (-iya)
Service
Shkola (-oly)
Sibirskiĭ (-skogo)
Skrifter
Slovak (-ovenská)
Society
Soobscheniya
Sovetskiĭ
Soviet
special
sperimentale
SSSR
Station
stiintifice
subtropicale
summary
Supplement
Survey
Symposium
System
Tagungsberichte
technical
Technology
Tekhnika
theoretical
thermal
Tidsskrift
Tijdschrift
Toxicology
Transactions
Travail (-aux)
tropical (-icale)
Trudy
Turkmenskaya
Ugeskrift
UK
Ukraĭnian
Ukrains'kaya

Breeding
British
Bulletin
California
Canada (-adian)
cellular (-ulaire)
Center
central
Centralblatt
Československý
chemical
Chemistry
chimicus
Chinese
Chromatography
Chronicle
Ciencia
cientificas
College
Commission
Communication
comparative
Comptes Rendus
Conference
Congress

Gruzinskaya
Helveticus
Histochemistry
Histoire (-ory)
Histology
horticultural
Horticulture
Hungaricae
Hungaricus
Husbandry
Hydrobiology
Hydrology
Indian
Industry
inorganic
Institute
Institutului
international
Investigation
Irrigation
Isotopes
Italian (-y)
Izvestiya
Jahrbuch
Japan (-anese)

Philosophy
Photogrammetric
Phycology
physical
Physics
physiological
Physiology
Phytologist
Phytopathology
Phytotaxonomy
Plantarum
Polonica (-ska)
Pollution
Práce
Practice
prikladnoĭ
Proceedings
Publication
Publishers
Quality
quantitative
Quarterly
Radiation
Radiobiology
Rasteniĭ

University
US, USA
USSR
Uzbekskiĭ (-ekskaya)
Vědecké (-ecký)
vegetable
végétale
Verhandlungen
Veröffentlichungen
Vestnik
Videnskabernes
Virology
Virusforschung
Viticulture
Volume
vsesoyuznyĭ
vyssheĭ (-iĭ)
výzkumný (-umného)
Weekblatt
Wetenschappen
Wissenschaft
Zeitschrift
Zeitung
Zentralblatt
Zhurnal

The numbers at the end of each reference of a journal article denote: volume (issue) : first page - last page, year of publication. The number of issue is given only in journal where each issue is paginated separately.

Book titles are cited according to the title page, not to the book jacket or cover. The publishing house, place and year of publication are included.

*2480 - ACEVES-N.E., STOLZY, L.H., MEHUYS, G.R.: Combined effects of oxygen and salinity on germination of a semi-dwarf mexican wheat. - Agron. J. 67: 530-532, 1975.

2481 - ACKERSON, R.C., KRIEG, D.R.: Stomatal and nonstomatal regulation of water use in cotton, corn, and sorghum. - Plant Physiol. 60: 850-853, 1977.

2482 - ACKERSON, R.C., KRIEG, D.R., HARING, C.L., CHANG, N.: Effects of plant water status on stomatal activity, photosynthesis, and nitrate reductase activity of field grown cotton. - Crop Sci. 17: 81-84, 1977.

2483 - ACKERSON, R.C., KRIEG, D.R., MILLER, T.D., STEVENS, R.G.: Water relations and physiological activity of potatoes. - J. Amer. Soc. hort. Sci. 102: 572-575, 1977.

2484 - ACKERSON, R.C., KRIEG, D.R., MILLER, T.D., ZARTMAN, R.E.: Water relations of field grown cotton and sorghum - temporal and diurnal changes in leaf water, osmotic, and turgor potentials. - Crop Sci. 17: 76-80, 1977.

2485 - ADAMS, J.A., JOHNSON, H.B., BINGHAM, F.T., YERMANOS, D.M.: Gaseous exchange of *Simmondsia chinensis* (Jojoba) measured with a double isotope porometer and related to water stress, salt stress and nitrogen deficiency. - Crop Sci. 17: 11-15, 1977.

2486 - AGARWAL, S.K., DE, R.: Effect of application of nitrogen, mulching and anti-transpirants on the growth and yield of barley under dryland conditions. - Ind. J. agr. Sci. 47: 191-194, 1977.

2487 - AHARONI, N., BLUMENFELD, A., RICHMOND, A.E.: Hormonal activity in detached lettuce leaves as affected by leaf water content. - Plant Physiol. 59: 1169-1173, 1977.

2488 - AHLGRIMM, H.-J.: Verschiedene Möglichkeiten der Feuchtegehaltsbestimmung. - Landbauforsch. Völkenrode 27: 97-104, 1977.

2489 - AHMAD, I., WAINWRIGHT, S.J.: Tolerance to salt, partial anaerobiosis, and osmotic stress in *Agrostis stolonifera*. - New Phytol. 79: 605-612, 1977.

2490 - AHO, N., DAUDET, F.A., VARTANIAN, N.: Régime transitoire de transpiration au cours de l'installation progressive d'une carence hydrique. - C.R. Acad. Sci. Paris, Sér. D 285: 159-162, 1977.

2491 - AKSENOV, S.I.: Ob otsenkakh sostoyaniya vody v biologicheskikh ob"ektakh po dannym razlichnykh fizicheskikh metodov. [On estimations of state of water in biological systems according to the data obtained by different physical methods] - Biofizika 22: 923-924, 1977. [In R, ab: E.]

2492 - AKSENOV, S.I., ASKOCHENSKAYA, N.A., GOLOVINA, E.A.: Izuchenie sostoyaniya vody v semenakh raznogo kachestvennogo sostava i ego izmeneniya pri temperaturnykh vozdeǐstviyakh. [The state of water studied in seeds of different qualitative composition and its changes upon the action of temperature.] - Fiziol. Rasl. 24: 1251-1260, 1977. [In R, ab: E.]

*2493 - AKSYONOV, S.I., GORYACHEV, S.N., FATEEVA, M.V., NIKITINA, T.N.: On the origin of slow components of NMR spin-echo decay in dried yeast and their correlation with yeast resistance to drying. - Studia biophys. 58: 121-129, 1976.

2494 - ALBERTE, R.S., THORNBER, J.P., FISCUS, E.L.: Water stress effects on the content and organization of chlorophyll in mesophyll and bundle sheath chloroplasts of maize. - Plant Physiol. 59: 351-353, 1977.

2495 - ALBRIGO, L.G.: Comparison of some antitranspirants on orange trees and fruit. - J. Amer. Soc. hort. Sci. 102: 270-273, 1977.

2496 - ALEKSEEVA, V.Ya., VELIKANOV, G.A., GORDON, L.Kh., BICHURINA, A.A.: Pronitsaemost' kletok korneǐ pshenitsy dlya vody pri indutsirovanii perekisnogo okisleniya lipidov. [Permeability of the cells of wheat roots for water during induction of lipid peroxidation.] - Fiziol. Rast. 24: 496-499, 1977. [In R, ab: E.]

2497 - ALESSI, J., POWER, J.F.: Residual effects of N-fertilazation on dryland spring wheat in Northern Plains. I. Wheat yield and water use. - Agron. J. 69: 1007-1011, 1977.

2498 - ALESSI, J., POWER, J.F., ZIMMERMAN, D.C.: Sunflower yield and water use as influenced by planting date, population, and row spacing. - Agron. J. 69: 465-469, 1977.

2499 - ALI, H.C.: Comparison of chlorophyll content and stomata size of inbred lines and their hybrids of corn (*Zea mays* L.). - Z. Acker- Pflanzenbau 145: 166 -170, 1977.

2500 - ALLEN, J.F.: Effects of washing and osmotic shock on catalase activity of intact chloroplast preparations. - FEBS Lett. 84: 221-224, 1977.

*2501 - ALLEN, L.H.,Jr., BOOTE, K.J., HAMMOND, L.C.: Peanut stomatal diffusion resistance affected by soil water and solar radiation. - Soil Crop Sci. Soc. Florida Proc. 35: 42-46, 1975.

*2502 - ALMADI, L.: Adatok az *Ambrosia elatior* vízháztartásához. [Water balance of *Ambrosia elatior*.] - Bot. Közlem. 63: 199-204, 1976. [In Hung, ab: E.]

2503 - AL-SAADI, H.A., WIEBE, H.H.: The influence of temperature and photoperiod on matric bound water in cabbage (*Brassica oleracea* L. var. capitata L.). - Z. Acker- Pflanzenbau 144: 205-208, 1977.

2504 - ALVAREZ, E.I., De DATTA, S.K.: Automatic feedback control to maintain constant soil moisture tension in the study of drought tolerance in rice. - Soil Sci. Soc. Amer. J. 41: 452-454, 1977.

2505 - AMIRDZHANOV, A.G.: Radiatsionnye faktory i transpiratsionnyĭ raskhod vinogradnika. [Radiation factors and transpiration expenditure of a vineyard.] - Fiziol. Rast. 24: 790-796, 1977. [In R, ab: E.]

2506 - ANDREWS, P., COLLINS, W.J., STERN, W.R.: The effect of withholding water during flowering on seed production in *Trifolium subterraneum* L. - Aust. J. agr. Res. 28: 301-307, 1977.

2507 - ANGUS, J.F., MONCUR, M.W.: Water stress and phenology in wheat. - Aust. J. agr. Res. 28: 177-181, 1977.

2508 - ANTIPOV, N.I.: Osobennosti vodoobmena gallov na list'yakh duba obyknovennogo. [Water exchange of galls on leaves of english oak.] - Fiziol. Rast. 24: 139 -142, 1977. [In R, ab: E.]

2509 - ANTIPOV, N.I.: O vodoobmene stebleĭ, chereskov i plastinok list'ev. [Water exchange of stems, stalks and blades of leaves.] - Fiziol. Rast. 24: 785-789, 1977. [In R, ab: E.]

2510 - ARDAKANI, M.S., FLÜHLER, H., McLAREN, A.D.: Rates of nitrate uptake with sudangrass and microbial reduction in a field. - Soil Sci. Soc. Amer. J. 41: 751-757, 1977.

2511 - ARIHARA, J., WATANABE, K.: [Method of water potential measurement and some results of experiment with that method.] - Jap. J. Crop Sci. 46: 137-141, 1977. [In Jap, ab: E.]

2512 - ARMSTRONG, R.A.: The response of lichen growth to additions of distilled water, rainwater and water from a rock surface. - New Phytol. 79: 373-376, 1977.

2513 - ARNOLD, W.N., LACY, J.S.: Permeability of the cell envelope and osmotic behaviour in *Saccharomyces cerevisiae*. - J. Bacteriol. 131: 564-571, 1977.

*2514 - ASHUROV, A.A., MOLOKOVSKIĬ, Yu.I.: Nekotorye anatomicheskie i fiziologicheskie osobennosti *Cotoneaster nummularis* Fisch. et Mey.(*Rosaceae*). [Some anatomical and physiological peculiarities of *Cotoneaster nummularis*.] - Bot. Zh. 61: 743-748, 1976. [In R.]

2515 - ASLAM, M., LOWE, S.B., HUNT, L.A.: Effect of leaf age on photosynthesis and transpiration of cassava (*Manihot esculenta*). - Can. J. Bot. 55: 2288-2295, 1977.

2516 - ASTHANA, D.C., VAMADEVAN, V.K.: Note on water losses in rice culture. - Ind. J. agr. Sci. 47: 216-217, 1977.

2517 - ATKINSON, R.G., ADAMSON, R.M.: Effects of fungicides on wilt and yield of greenhouse tomatoes grown in *Fusarium*-infested sawdust. - Can. J. Plant Sci. 57: 675-680, 1977.

*2518 - AUCLAIR, D.: Effets des poussières sur la photosynthèse. I. Effets des poussères de ciment et de charbon sur la photosynthèse de l'épicéa. - Ann. Sci. forest. 33: 247-255, 1976.

2519 - AYERBE, L., GOMEZ-CAMPO, C.: Drought hardening in excised shoots of *Bryophyllum tubiflorum* Harv. - Ann. Bot. 41: 649-651, 1977.

2520 - AYERS, R.S.: Quality of water for irrigation. - J. Irrig. Drain Div. - ASCE 103: 135-154, 1977.

2521 - AYRES, P.G.: Effect of water potential of pea leaves on spore production by *Erysiphe pisi* (powdery mildew). - Trans. Brit. mycol. Soc. 68: 97-100, 1977.

2522 - AYRES, P.G.: Effects of powdery mildew *Erysiphe pisi* and water stress upon the water relations of pea. - Physiol. Plant Pathol. 10: 139-145, 1977.

2523 - BAJPAI, M.R., MERTIA, H.S.: A note on the irrigation and fertility levels on the yield and nutrient uptake of dwarf barley cultivar RDB-1. - Ann. Arid Zone 16: 153-156, 1977.

2524 - BAKER, C.J.: Some effects of cover, seed size, and soil moisture status on establishment  of seedlings by direct drilling. - N. Zeal. J. exp. Agr. 5: 47-53, 1977.

*2525 - BALLARD, L.A.T.: Strophiolar water conduction in seeds of the *Trifolieae* induced by action on the testa at non-strophiolar sites. - Aust. J. Plant Physiol. 3: 465-469, 1976.

*2526 - BALLIO, A., D'ALESSIO, V., RANDAZZO, G., BOTTALICO, A., GRANITI, A., SPARAPANO, L., BOSNAR, B., CASINOVI, C.G., GRIBANOVSKI-SASSU, O.: Occurence of fusicoccin in plant tissues infected by *Fusicoccum amygdali* Del. - Physiol. Plant Pathol. 8: 163-169, 1976.

*2527 - BAŇOCH, Z., HÁJEK, L.: Funkce živin při stupňované agrotechnice včetně závlahy u kukuřice na zrno. [The function of nutrients under different cultural practices including irrigation in grain maize.] - Rost. Výroba (Praha) 22: 1029-1039, 1976. [In Czech, ab: E,R.]

2528 - BARA, M.: The effect of horizontal clinostat on growth, water and protein contents in *Helianthus annuus* hypocotyls. - In: BOGORAD, L., WEIL, J.H. (ed.): Acides Nucléiques et Synthèse des Protéines chez les Végétaux. Coll. Int. C.N.R.S. No. 261. Pp. 629-633. Édit. CNRS, Paris 1977.

*2529 - BARADAS, M.W., BLAD, B.L., ROSENBERG, N.J.: Reflectant induced modification of soybean canopy radiation balance. V. Longwave radiation balance. - Agron. J. 68: 848-852, 1976.

*2530 - BARKHAM, J.P., RAINY, M.E.: The vegetation of the Samburu-Isiolo Game Reserve. - East Afr. Wildl. J. 14: 297-329, 1976.

2531 - BARLOW, E.W.R., BOERSMA, L., YOUNG, J.L.: Photosynthesis, transpiration, and leaf elongation in corn seedlings at suboptimal soil temperatures. - Agron. J. 69: 95-100, 1977.

2532 - BARLOW, E.W.R., MUNNS, R., SCOTT, N.S., REISNER, A.H.: Water potential, growth, and polyribosome content of stressed wheat apex. - J. exp. Bot. 28: 909-916, 1977.

*2533 - BARRADA, Y., HALSTEAD, E., NETHSINGHE, D.A.: Fertilizer and water efficiency studies. - In: Efficiency of Water and Fertilizer Use in Semi-Arid Regions. A Technical Document. Pp. 135-158. International Atomic Energy Agency, Vienna 1976.

*2534 - BARTHE, P., Le PAGE-DEGIVRY, M.-T., BULARD, C.: Mise au point d'un protocole expérimental permettant d'évaluer quantitativement l'acide trans-abscissique à l'aide de tests biologiques. - C.R. Acad. Sci. Paris, Sér. D 283: 1297-1299, 1976.

2535 - BAR-YOSEF, B.: Trickle irrigation and fertilization of tomatoes in sand dunes: water, N, and P distribution in the soil and uptake by plants. - Agron. J. 69: 486-491, 1977.

2536 - BASSIRI, A., KHOSH-KHUI, M., ROUHANI, I.: The influences of simulated moisture stress conditions and osmotic substrates on germination and growth of cultivated and wild safflowers. - J. agr. Sci. 88: 95-100, 1977.

*2537 - BAUER, A., CASSEL, D.K., ZIMMERMAN, L.: Fertilizer effects on corn production under irrigation at Oakes. - North Dakota Agr. Exp. Sta. Res. Rep. 60: 1-24, 1975.

*2538 - BAUER, A., HEIMBUCH, T., CASSEL, D.K., ZIMMERMAN, L.: Production potential of sugarbeets under irrigation in the West Oakes irrigation district. - North Dakota Agr. Exp. Sta. Bull. 498: 1-94, 1975.

2539 - BAUR, J.R., MILLER, P.R., BOVEY, R.W.: Effects of preharvest desiccation with glyphosate on grain sorghum seed. - Agron. J. 69: 1015-1018, 1977.

2540 - BAXTER, P., WEST, D.: The flow of water into fruit trees. I. Resistances to water flow through roots and stems. - Ann. appl. Biol. 87: 95-101, 1977.

2541 - BEADLE, C.L., JARVIS, P.G.: The effects of shoot water status on some photosynthetic partial processes in Sitka spruce. - Physiol. Plant. 41: 7-13, 1977.

2542 - BECKER, M.: Contribution à l'etude de la transpiration et de l'adaptation à la sécheresse des jeunes plant résineux. Exemple de 3 sapins du pourtour méditerranéen (*Abies alba, A. Nordmanniana, A. numidica*). - Ann. Sci. forest. 34: 137-158, 1977.

2543 - BECWAR, M.R., MANSOUR, N.S., VARSEVELD, G.W.: Microwave drying: a rapid method for determining sweet corn moisture. - HortScience 12: 562-563, 1977.

2544 - BEECH, D.F.: Growth and oil production of lemongrass (*Cymbopogon citratus*) in Ord irrigation area, Western Australia. - Aust. J. exp. Agr. anim. Husb. 17: 301-307, 1977.

2545 - BEHBOUDIAN, M.H.: Water relations of cucumber, tomato, and sweet pepper. - Med. Landbouwhogesch. Wageningen 77-6: 1-84, 1977.

2546 - BEHBOUDIAN, M.H., HOLSTEIJN, H.M.C., van: Water relations of lettuce. I. Internal physical aspects for two cultivars. - Scientia Hort. 7: 9-17, 1977.

2547 - BELETSKAYA, E.K.: Izmenenie metabolizma ozimykh kul'tur pri ikh adaptatsii k zatopleniyu. [Changes in metabolism of winter cultures during their adaptation to flooding.] - Fiziol. Rast. 24: 924-932, 1977. [In R, ab: E.]

2548 - BENEDETTI, E., de, GARLASCHI, F.M.: On the estimation of proton gradient and osmotic volume in chloroplast membranes. - J. Bioenerg. Biomembranes 9: 195 -201, 1977.

2549 - BENGTSON, C., FALK, S.O., LARSSON, S.: The after-effect of water stress on transpiration rate and changes in abscisic acid content of young wheat plants. - Physiol. Plant. 41: 149-154, 1977.

2550 - BERGMANN, H.: Möglichkeiten der Transpirationseffektuierung und Stabilisierung des Pflanzenwasserhaushaltes unter besonderer Berücksichtigung einer Agrochemikalienanwendung. Übersichtsbeitrag. - Arch. Acker- Pflanzenbau Bodenk. 21: 767-788, 1977.

2551 - BHAGSARI, A.S., ASHLEY, D.A., BROWN, R.H., BOERMA, H.R.: Leaf photosynthetic characteristics of determinate soybean cultivars. - Crop Sci. 17: 929-932, 1977.

*2552 - BHANDARI, M.C., SEN, D.N.: Interaction of light, potassium chloride and potassium cyanide in the stomatal regulation of *Citrullus colocynthis* (Linn.) Schard. - Botanique 7: 21-28, 1976.

*2553 - BHATT, J.G., APPUKUTTAN, E.: Soil plant relationships in irrigated cottons. - Mysore J. agr. Sci. 10: 522-529, 1976.

2554 - BIEDERBECK, V.O., CAMPBELL, C.A., NICHOLAICHUK, W.: Simulated dew formation and microbial growth in soil of a semiarid region of western Canada. - Can. J. Soil Sci. 57: 93-102, 1977.

2555 - BILAN, M.V., HOGAN, C.T., CARTER, H.B.: Stomatal opening, transpiration, and needle moisture in loblolly pine seedlings from two Texas seed sources. - Forest Sci. 23: 457-462, 1977.

2556 - BINET, P., TREMBLIN, G.: Influence d'une secheresse edaphique simulee (emploi de PEG 600) sur la croissance et la composition minerale du mais. - Bull. Soc. Linn. Normandie 105: 117-122, 1977.

2557 - BINGHAM, G.E., COYNE, P.I.: A portable, temperature-controlled, steady-state porometer for field measurement of transpiration and photosynthesis. - Photosynthetica 11: 148-160, 1977.

2558 - BISCOE, P.V., INCOLL, L.D., LITTLETON, E.J., OLLERENSHAW, J.H.: Barley and its environment. VII. Relationships between irradiance, leaf photosynthetic rate and stomatal conductance. - J. appl. Ecol. 14: 293-302, 1977.

2559 - BLACK, J.D.F., MITCHELL, P.D., NEWGREEN, P.N.: Optimum irrigation rates for young trickle irrigated peach trees. - Aust. J. exp. Agr. anim. Husb. 17: 342-345, 1977.

2560 - BLAKE, J., FERRELL, W.K.: The association between soil and xylem water potential, leaf resistance, and abscisic acid content in droughted seedlings of Douglas-fir (*Pseudotsuga menziesii*). - Physiol. Plant. 39: 106-109, 1977.

2561 - BLAKE, M.I., UPHAUS, R.A., KATZ, J.J.: The effects of heavy water on higher plants: a review. - Lloydia 40: 127-135, 1977.

2562 - BLAKE, T.J.: Diurnal rhythms in stem elongation in *Eucalyptus obliqua* in relation to temperature and seedling water status. - Aust. J. Bot. 25: 455-459, 1977.

2563 - BLANCHET, R., GELFI, N., BOSC, M.: Relations entre consommation d'eau et production chez divers types variétaux de soja ( *Glycine max*. L. Merr.). - Ann. agron. 28: 261-275, 1977.

2564 - BLEKHMAN, G.I.: Vozmozhnyĭ mekhanizm izmeneniya tsitoplazmaticheskoĭ ribonukleaznoĭ aktivnosti v list'yakh prorostkov pshenitsy pri obezvozhivanii i regidratatsii. [Possible mechanism involved in changes of cytoplasmic ribonuclease activity in leaves of wheat seedlings during dehydration and rehydration.] - Fiziol. Rast. 24: 507-512, 1977. [In R, ab: E.]

2565 - BOARDMAN, N.K.: Comparative photosynthesis of sun and shade plants. - Annu. Rev. Plant Physiol. 28: 355-377, 1977.

2566 - BOARDMAN, N.K.: Regulation of photosynthesis by light intensity. - Proc. Aust. biochem. Soc. 10: Q8, 1977.

2567 - BOBROVSKAYA, N.I., ZAVADSKAYA, I.G., KOBAK, K.I.: Vliyanie razlichnoĭ stepeni degidratatsii na dvizhenie protoplazmy, sposobnost' k plazmolizu i dykhanie u *Cleistogenes squarrosa* (Trin.) Keng. i *Agropyron cristatum* (L.) Beauv. [Influence of different water deficits on streaming of protoplasma, ability to plasmolysis and respiration in *Cleistogenes squarrosa* and *Agropyron cristatum*.] - In: Problemy Ékologii, Geobotaniki, Botanicheskoĭ Geografii i Floristiki. Pp. 201-206. Nauka, Leningrad 1977. [In R.]

2568 - BÖCHER, M., KLUGE, M.: Der $C_4$-Weg der C-Fixierung bei *Spinacea oleracea*. I. $^{14}$C-Markierungsmuster suspendierter Blattstreifen unter dem Einfluss des Suspensionsmedium. - Z. Pflanzenphysiol. 83: 347-361, 1977.

2569 - BOGGIE, R.: Water-table depth and oxygen content of deep peat in relation to root growth of *Pinus contorta*. - Plant Soil 48: 447-454, 1977.

2570 - BOGGIE, R.: A simple device for recording maximum and minimum water-table levels in soils. - J. appl. Ecol. 14: 283-285, 1977.

2571 - BOĬKO, G.N.: Deĭstvie razlichnykh osmoticheski aktivnykh veshchestv na rost zarodysheĭ yabloni v steril'noĭ kul'ture. [Effect of different osmotically active substances on growth of apple embryos in sterile culture.] - Fiziol. Biokhim. kul't. Rast. 9· 637-641, 1977. [In R, ab: E.]

2572 - BOLE, J.B.: Uptake of $^3$HHO and $^{32}$P by roots of wheat and rape. - Plant Soil 46: 297-307, 1977.

*2573 - BONHOMME, R., VARLET GRANCHER, C.: Assimilation nette, utilisation de l'eau et microclimatologie d'un champ de mais. VII. Variations du rapport des énergies rouge sombre-rouge clair. - Ann. agron. 27: 327-332, 1976

2574 - BONTE, J., BONTE, C., CORMIS, L., de, LOUGUET, P.: Contribution à l'étude des caractères de résistance de *Pelargonium* à un polluant atmosphérique, le dioxyde de soufre. - Physiol. vég. 15: 15-27, 1977.

2575 - BONTE, J., CORMIS, L., de, LOUGUET, P.: Inhibition, en anaerobiose, de la reaction de fermeture des stomates du *Pelargonium* en presence de dioxyde de soufre. - Environ. Pollut. 12: 125-133, 1977.

*2576 - BOOTE, K.J., VARNELL, R.J., DUNCAN, W.G.: Relationships of size, osmotic concentration, and sugar concentration of peanut pods to soil water. - Soil Crop Sci. Soc. Florida Proc. 35: 47-50, 1975.

*2577 - BORNMAN, C.H., FANSHAWE, N.C.: *Welwitschia mirabilis* callus studies. IV. Some effects of stress factors. - Z. Pflanzenphysiol. 78: 334-338, 1976.

*2578 - BOTMAN, K.S., GIRSHEVICH, E.I.: Raskhod vody na transpiratsiyu orekhovymi drevostoyami razlichnoi somknutosti pologa. [Transpirational water efflux by *Juglans regia* stand of different density.] - Uzb. biol. Zh. 1976(4): 27-30, 1976. [In R.]

2579 - BOURZEIX, M., HEREDIA, N., MERIAUX, S., ROLLIN, H., RUTTEN, P.: De l'influence de l'alimentation hydrique de la Vigne sur les caractéristiques anatomiques des Baies de Raisins et leur richesse en couleur, tanins et autres constituants phénoliques. - C.R. Acad. Sci. Paris, Sér. D 284: 365-368, 1977.

*2580 - BOUSSIBA, S., RIKIN, A., RICHMOND, A.E.: The role of abscisic acid in cross-adaptation of tobacco plants. - Plant Physiol. 56: 337-339, 1975.

2581 - BOYER, J.S.: Regulation of water movement in whole plants. - In: JENNINGS, D.H. (ed.): Integration of Activity in the Higher Plant. Pp. 445-470. Cambridge University Press, Cambridge - London - New York - Melbourne 1977.

2582 - BRADBURY, I.K., MALCOLM, D.C.: The effect of phosphorus and potassium on transpiration, leaf diffusive resistance and water-use efficiency in Sitka spruce (*Picea sitchensis*) seedlings. - J. appl. Ecol. 14: 631-641, 1977.

2583 - BRANDLE, J.R., HINCKLEY, T.M., BROWN, G.N.: The effects of dehydration-rehydration cycles on protein synthesis of black locust seedlings. - Physiol. Plant. 40: 1-5, 1977.

2584 - BRAR, G., THIES, W.: Contribution of leaves, stem, siliques and seeds to dry matter accumulation in ripening seeds of rapeseed, *Brassica napus* L. - Z. Pflanzenphysiol. 82: 1-13, 1977.

2585 - BRAUN, G.: Über die Ursachen und Kriterien der Immissionsresistenz bei Fichte *Picea abies* (L.) Karst. I. Morphologisch-anatomische Immissionsresistenz. - Europ. J. Forest Pathol. 7: 23-43, 1977.

*2586 - BRAUN, H.J.: Rhytmus und Grösse von Wachstum, Wasserverbrauch und Produktivität des Wasserverbrauchs bei Holzpflanzen. II. *Acer platanoides* L., *Acer pseudoplatanus* L. und *Fraxinus excelsior* L. mit einem Vergleich aller untersuchter Baumarten einschliesslich einiger *Populus*-Klone. - Allg. Forst- Jagdzeit. 147: 163-168, 1977.

2587 - BRAUN, H.J.: Das Wuchsverhalten einiger bekannter Herkünfte der Fichte (*Picea abies* (L.) Karst.) in unterschiedlich belüftetem Bodenmilieu. - Allg. Forst- Jagdzeit. 148: 1-3, 1977.

2588 - BRAUN, H.J.: Zum Wachstum und zur Produktivität des Wasserverbrauches der Baumarten *Acer platanoides* L., *Acer pseudoplatanus* L. und *Fraxinus excelsior* L. - Z. Pflanzenphysiol. 84: 459-462, 1977.

2589 - BRAVDO, B.-A.: Oscillatory transpiration and $CO_2$ exchange of citrus leaves at the $CO_2$ compensation concentration. - Physiol. Plant. 41: 36-41, 1977.

*2590 - BROCKWELL, J., GAULT, R.R.: Effects of irrigation water temperature on growth of some legume species in glasshouses. - Aust. J. exp. Agr. anim. Husb. 16: 500-505, 1976.

*2591 - BROUÉ, P., MARSHALL, D.R., MUNDAY, J.: The response of lupins to waterlogging. - Aust. J. exp. Agr. anim. Husb. 16: 549-554, 1976.

2592 - BROWN, L.F., TRLICA, M.J.: Interacting effects of soil water, temperature and irradiance on $CO_2$ exchange rates of two dominant grasses of the shortgrass prairie. - J. appl. Ecol. 14: 197-204, 1977.

2593 - BROWN, L.F., TRLICA, M.J.: Carbon dioxide exchange of blue grama swards as influenced by several ecological variables in the field. - J. appl. Ecol. 14: 205-213, 1977.

2594 - BROWN, M.J., WRIGHT, J.L., KOHL, R.A.: Onion-seed yield and quality as affected by irrigation management. - Agron. J. 69: 369-372, 1977.

2595 - BROWN, R.W., McDONOUGH, W.T.: Thermocouple psychrometer for in situ leaf water potential determinations. - Plant Soil 48: 5-10, 1977.

2596 - BROWN, T.H.: Rate of loss of dry matter and change in chemical composition of of nine pasture species over summer. - Aust. J. exp. Agr. anim. Husb. 17: 75 -79, 1977.

2597 - BROWNE, C.L.: Effect of date of final irrigation on yield and yield components of sunflowers in a semi-arid environment. - Aust. J. exp. Agr. anim. Husb. 17: 482-488, 1977.

2598 - BROWNING, G.: Environmental control of flower bud development in *Coffea arabica* L. - In: LANDSBERG, J.J., CUTTING, C.V.(ed.): Environmental Effects on Crop Physiology. Pp. 321-336. Academic Press, London - New York - San Francisco 1977.

2599 - BRUNNER, U., ELLER, B.M.: Spectral properties of juvenile and adult leaves of *Piper betle* and their ecological significance. - Physiol. Plant. 41: 22-24, 1977.

*2600 - BRZOZOWSKA, J., HANOWER, P.: Sur les composes phenoliques des vegetaux et leur rapport avec un deficit hydrique chez des cotonniers. - Ann. Univ. Abidjan, Sér. C 12: 65-87, 1976.

2601 - BUCHANAN, B.A., DAVAULT, M.F., FISHER, J.T.: Influence of artificial shade on water stress of containerized ponderosa pine seedlings. - Can. J. Forest Res. 7: 537-540, 1977.

2602 - BUCKNER, E., MAKI, T.E.: Seven-year growth of fertilized and irrigated yellow -poplar, sweetgum, northern red oak, and loblolly pine planted on two sites. - Forest. Sci. 23: 402-410, 1977.

2603 - BUNCE, J.A.: Nonstomatal inhibition of photosynthesis at low water potentials in intact leaves of species from a variety of habitats. - Plant Physiol. 59: 348-350, 1977.

2604 - BUNCE, J.A.: Leaf elongation in relation to leaf water potential in soybean. - J. exp. Bot. 28: 156-161, 1977.

2605 - BUNCE, J.A., MILLER, L.N., CHABOT, B.F.: Competitive exploitation of soil water by five eastern North American tree species. - Bot. Gaz. 138: 168-173, 1977.

2606 - BUNCE, J.A., PATTERSON, D.T., PEET, M.M., ALBERTE, R.S.: Light acclimation during and after leaf expansion in soybean. - Plant Physiol. 60: 255-258, 1977.

*2607 - BURDEN, R.S., TAYLOR, H.F.: Xanthosin and abscisic acid. - Pure appl. Chem. 47: 203-209, 1976.

2608 - BURGE, M.N., ISAAC, I.: Predisposition of *Aster* to *Phialophora* wilt. - Ann. appl. Biol. 86: 353-358, 1977.

2609 - BUTCHER, T.B.: Impact of moisture relationships on the management of *Pinus pinaster* Ait. plantations in Western Australia. - Forest Ecol. Management 1: 97-107, 1977.

*2610 - BUTCHER, T.B., HAVEL, J.J.: Influence of moisture relationships on thinning practice. - N. Zeal. J. Forest. Sci. 6: 158-170, 1976.

2611 - BUTLER, D.R.: Coffee leaf temperatures in a tropical environment. - Acta Bot. Neerl. 26: 129-140, 1977.

2612 - BYRNE, G.F., BEGG, J.E., HANSEN, G.K.: Cavitation and resistance to water flow in plant roots. - Agr. Meteorol. 18: 21-25, 1977.

2613 - BYSZEWSKI, W.: Zusammenhang zwischen der Produktivität und der Ertragsfähigkeit von Zuckerrüben. - In: Produkce Biomasy a Tvorba Výnosu Polních Plodin. Vol. 1. Pp. 115-135. ČVTSZ, Praha 1977.

2614 - CA..WELL, M.M., WHITE, R.S., MOORE, R.T., CAMP, L.B.: Carbon balance, productivity, and water use of cold-winter desert shrub communities dominated by $C_3$ and $C_4$ species. - Oecologia 29: 275-300, 1977.

*2615 - CAMPBELL, C.A., BIEDERBECK, V.O.: Soil bacterial changes as affected by growing season weather conditions: a field and laboratoty study. - Can. J. Soil Sci. 56: 293-310, 1976.

2616 - CAMPBELL, C.A., CAMERON, D.R., NICHOLAICHUK, W., DAVIDSON, H.R.: Effect of fertilizer N and soil moisture on growth, N content, and moisture use by spring wheat. - Can. J. Soil Sci. 57: 289-310, 1977.

2617 - CAMPBELL, C.A., DAVIDSON, H.R., WARDER, F.G.: Effects of fertilizer N and soil moisture on yield, yield components, protein content and N accumulation in the aboveground parts of spring wheat. - Can. J. Soil Sci. 57: 311-327, 1977.

2618 - CAMPBELL, G.S.: An Introduction to Environmental Biophysics. - Springer-Verlag, New York - Heidelberg - Berlin 1977.

2619 - CAMPBELL, G.S., HARRIS, G.A.: Water relations and water use patterns for *Artemisia tridentata* Nutt. in wet and dry years. - Ecology 58: 652-659, 1977.

2620 - CAMPBELL, R.B., PHENE, C.J.: Tillage, matric potential, oxygen and millet yield relations in a layered soil. - Trans. ASAE 20: 271-275, 1977.

2621 - CARLIER, G., BOULENS, D., LABROT, S.: Relations entre le potentiel hydrique et l'aptitude à l'absorption de 3-O-méthyl-D-glucose dans les tissus foliaires de *Pelargonium zonale* (L.) Aiton. - Physiol. Vég. 15: 695-703, 1977.

2622 - CARLSON, R.W., BAZZAZ, F.A.: Growth reductions in American sycamore (*Plantanus occidentalis* L.) caused by Pb-Cd interaction. - Environ. Pollut. 12: 243-253, 1977.

2623 - CARR, M.K.V.: Responses of seedling tea bushes and their clones to water stress. - Exp. Agr. 13: 317-324, 1977.

2624 - CARREKER, J.R., WILKINSON, S.R., BARNETT, A.P., BOX, J.E.: Soil and water management systems for sloping land. - Agr. Res. Serv. Publ. S-160: 1-76, 1977.

2625 - CARVER, T.L.W., CARR, A.J.H.: Race non-specific resistance of oats to primary infection by mildew. - Ann. appl. Biol. 86: 29-36, 1977.

2626 - CARY, J.W.: Photosynthesis of sugarbeets under N and P stress: field measurements and carbon balance. - Agron. J. 69: 739-744, 1977.

2627 - CASTRO, P.R.C., MALAVOLTA, E.: Influence of growth regulators upon mineral nutrition, osmotic potential, and incidence of blossom-end rot of tomato fruit. - Turrialba 27: 273-276, 1977.

2628 - CASTRO, P.R.C., MALAVOLTA, E., MORAES, R.S.: Potential osmótico foliar de tomateiros sob efeito de reguladores de crescimento. [The effect of growth substances on the osmotic potential of tomato leaves.] - Rev. Brasil. Biol. 37: 785-789, 1977. [In Span, ab: E.]

2629 - CERDA, A., BINGHAM, F.T., HOFFMAN, G.J.: Interactive effect of salinity and phosphorus on sesame. - Soil. Sci. Soc. Amer. J. 41: 915-918, 1977.

2630 - ČERMÁK, J.: Průběh transpiračního proudu u vzrostlých stromů z hlediska různých časových měřítek. [The course of transpiration flow rate in full grown trees in different time scales.] - In: HUZULÁK, J., MAŠAROVIČOVÁ, E. (ed.): Fotosyntéza a Vodný Režim Drevín. Pp. 47-54. Modra-Piesky 1977. [In Czech, ab: E.]

*2631 - CERNUSCA, A.: Bestandesstruktur, Bioklima und Energiehaushalt von alpinen Zwergstrauchbeständen. - Ecol. Plant. 11: 71-102, 1976.

*2632 - CERNUSCA, A.: Energie- und Wasserhaushalt eines alpinen Zwergstrauchbestandes während einer Föhnperiode. - Arch. Meteorol. Geophys. Bioklimatol., Ser. B 24: 219-241, 1976.

2633 - CERNUSCA, A.: Bestandesstruktur, Mikroklima, Bestandesklima und Energiehaushalt von Pflanzenbeständen des alpinen Grasheidegürtels in den Hohen Tauern. Erste Ergebnisse der Projektstudie 1976. - Veröff. Österreich. Mass-Hochgebirgsprogrammes Hohe Tauern 1: 25-45, 1977.

*2634 - CERNUSCA, A., CERNUSCA, G.: Ein automatischer Wasserstands-Niveaugeber für die elektrische Messung von Verdunstung, Evapotranspiration oder Niederschlag. - Wetter Leben 28: 28-33, 1976.

2635 - CHABOT, J.F., CHABOT, B.F.: Ultrastructure of the epidermis and stomatal complex of balsam fir (*Abies balsamea*). - Can. J. Bot. 55: 1064-1075, 1977.

*2636 - CHALLA, H.: An analysis of the diurnal course of growth, carbon dioxide exchange and carbohydrate reserve content of cucumber. - Agr. Res. Rep. (Wageningen) 861: 1-88, 1976.

2637 - CHANEY, W.R., KOZLOWSKI, T.T.: Patterns of water movement in intact and excised stems of *Fraxinus americana* and *Acer saccharum* seedlings. - Ann. Bot. 41: 1093-1100, 1977.

*2638 - CHANNAPPA, T.C.: Prediction of potential evapotranspiration. - Mysore. J. agr. Sci. 10: 83-98, 1976.

2639 - CHAPMAN, D.C., RAND, R.H., COOKE, J.R.: A hydrodynamical model of bordered pits in conifer tracheids. - J. theor. Biol. 67: 11-24, 1977.

2640 - CHATURVEDI, S.N., ZABKA, G.: Studies on dark fixation of carbon dioxide in *Kalanchöe*. I. Effect of water stress and growth retardants. - Ann. Bot. 41: 493-500, 1977.

2641 - CHATURVEDI, S.N., ZABKA, G.: Studies on dark fixation of carbon dioxide in *Kalanchöe*. II. Effect of interaction of photoperiodic induction with water stress and growth retardants. - Ann. Bot. 41: 501-505, 1977.

2642 - CHEN, P.M., LI, P.H.: Induction of frost hardiness in stem cortical tissues of *Cornus stolonifera* Michx. by water stress. II. Biochemical changes. - Plant Physiol. 59: 240-243, 1977.

2643 - CHEN, P.M., LI, P.H., BURKE, M.J.: Induction of frost hardiness in stem cortical tissues of *Cornus stolonifera* Michx. by water stress. I. Unfrozen water in cortical tissues and water status in plants and soil. - Plant Physiol. 59: 236 -239, 1977.

2644 - CHESNESS, J.L., HENDERSHOTT, C.H., COUVILLON, G.A.: Evaporative cooling of peach trees to delay bloom. - Trans. ASAE 20: 466-468, 1977.

*2645 - CHETAL, S., NAINAWATEE, H.S.: Some enzyme changes associated with water-stress in germinating paddy and barley seeds. - Plant biochem. J. 3: 105-110, 1976.

2646 - CHILDS, S.W., GILLEY, J.R., SPLINTER, W.E.: A simplified model of corn growth under moisture stress. - Trans. ASAE 20: 858-865, 1977.

2647 - CHIN CHOY, E.W., STONE, J.F., GARTON, J.E.: Row spacing and direction effects on water uptake characteristics of peanuts. - Soil Sci. Soc. Amer. J. 41: 428-432, 1977.

2648 - CHHINNAN, M.S., YOUNG, J.H.: A study of diffusion equations describing moisture movement in peanut pods - II: Simultaneous vapor and liquid diffusion of moisture. - Trans. ASAE 20: 749-753, 757, 1977.

2649 - CHINNAN, M.S., YOUNG, J.H.: A study of diffusion equations describing moisture movement in peanut pods - I. Comparison of vapor and liquid diffusion equations. - Trans. ASAE 20: 539-546, 1977.

2650 - CHOLICK, F.A., WELSH, J.R., COLE, C.V.: Rooting patterns of semi-dwarf and tall winter wheat cultivars under dryland field conditions. - Crop Sci. 17: 637-639, 1977.

2651 - CHOW, T.L.: A porous cup soil-water sampler with volume control. - Soil Sci. 124: 173-176, 1977.

*2652 - CHRISTERSSON, L.: The effect of inorganic nutrients on water economy and hardiness of conifers. II. The effect of varying potassium and calcium contents on water status and drought hardiness of pot-grown *Pinus silvestris* L. and *Picea abies* (L.) Karts. seedlings. - Studia Forest. Suecica 136: 1-23, 1976.

2653 - CHRISTIANSEN, J.E., OLSEN, E.C., WILLARDSON, L.S.: Irrigation water quality evaluation. - J. Irrig. Drain. Div. ASCE 103: 155-169, 1977.

*2654 - CHRISTY, A.L.: Mathematical models of Münch pressure flow: basic concepts and assumptions. - In: WARDLAW, J.F., PASSIOURA, J.B. (ed.): Transport and Transfer Processes in Plants. Pp. 363-368. Academic Press, New York - San Francisco - London 1976.

2655 - CHU, A.C.P., McPHERSON, H.G.: Sensitivity to desiccation of leaf extension in prairie grass. - Aust. J. Plant Physiol. 4: 381-388, 1977.

*2656 - CHU, A.C.P., TILLMAN, R.F.: Growth of a forage sorghum hybrid under two soil moisture regimes in the Manawatu. - N.Zeal. J. exp. Agr. 4: 351-355, 1976.

2657 - CLARK, B.J., PRIOUL, J.-L., COUDERC, H.: The physiological response to cutting in Italian ryegrass. - J. Brit. Grassland Soc. 32: 1-5, 1977.

2658 - CLAUS, S., UNGER, K.: Energie- und Wasserhaushalt eines Blattes unter Berücksichtigung der Stomataregelung. - In: UNGER, K.(ed.): Biophysikalische Analyse Pflanzlicher Systeme. Pp. 131-139. VEB Gustav Fischer Verlag, Jena 1977.

2659 - CLEMENS, J., PEARSON, C.J.: The effect of waterlogging on the growth and ethylene content of *Eucalyptus robusta* Sm. (Swamp Mahogany). - Oecologia 29: 249-255, 1977.

2660 - CLINE, J.F., URESK, D.W., RICKARD, W.H.: Comparison of soil water used by a sagebrush-bunchgrass and cheatgrass community. - J. Range Manage. 30: 199-201, 1977.

2661 - COHEN, Y.: The combined effects of temperature, leaf wetness, and inoculum concentration on infection of cucumbers with *Pseudoperonospora cubensis*. - Can. J. Bot. 55: 1478-1487, 1977.

2662 - COLE, N.H.A.: Effect of light, temperature, and flooding on seed germination of the neotropical *Panicum laxum* Sw. - Biotropica 9: 191-194, 1977.

2663 - COLLATZ, G.J.: Influence of certain environmental factors on photosynthesis and photorespiration in *Simmondsia chinensis*. - Planta 134: 127-132, 1977.

2664 - COLLINS, O.D.G., SUTCLIFFE, J.F.: The relationship between transport of individual elements and dry matter from the cotyledons of *Pisum sativum* L. - Ann. Bot. 41: 163-171, 1977.

2665 - CONNOR, D.J., LEGGE, N.J., TURNER, N.C.: Water relations of mountain ash (*Eucalyptus regnans* F. Muell.) forests. - Aust. J. Plant Physiol. 4: 753-762, 1977.

2666 - COOK, E.R., JACOBY, G.C., Jr.: Tree-ring-drought relationships in the Hudson Valley, New York. - Science 198: 399-401, 1977.

*2667 - COOKE, R.J., De BAERDEMAEKER, J.G., RAND, R.H., MANG, H.A.: A finite element shell analysis of guard cell deformations. - Trans. ASAE 19: 1107-1121, 1976.

*2668 - CORLEY, R.V.H., HARDON, J.J., WOOD, B.J.: Developments in Crop Science (1). Oil Palm Research. - Elsevier Scientific Publication Company, Amsterdam - New York 1976.

2669 - CORNILLON, P.: Influence de la température du substrat sur la composition des racines et des limbes de tomate. Implications concernant l'absorption hydrique. - Ann. agron. 28: 277-289, 1977.

2670 - CORNISH, P.S., MYERS, L.F.: Low pasture productivity of a sedimentary soil in relation to phosphate and water supply. - Aust. J. exp. Agr. anim. Husb. 17: 776-783, 1977.

*2671 - COSTER, H.G.L., STEUDLE, E., ZIMMERMANN, U.: Turgor pressure sensing in plant cell membranes. - Plant Physiol. 58: 636-643, 1976.

2672 - COUCHAT, P.: Aspects méthodologiques et technologiques de la mesure neutronique de l'humidité des sols. - Ann. agron. 28: 477-488, 1977.

2673 - COUCHAT, P.: Effet de l'oxygène sur la transpiration. - C.R. Acad. Sci. Paris, Sér. D 285: 1303-1306, 1977.

2674 - COUDRET, A., FERRON, F.: La transpiration végétale. Modes d'action des antitranspirants. - Ann. Amélior. Plant. 27: 613-638, 1977.

2675 - COUTTS, M.P.: The formation of dry zones in the sapwood of conifers. II. The role of living cells in the release of water.-Europ. J. Forest Pathol. 7: 6-12, 1977.

2676 - COUTTS, M.P., RISHBETH, J.: The formation of wetwood in Grand fir.. - Europ. J. Forest. Pathol. 7: 13-22, 1977.

2677 - COWAN, I.R., FARQUHAR, G.D.: Stomatal function in relation to leaf metabolism and environment. - In: JENNINGS, D.H. (ed.): Integration of Activity in the Higher Plant. Pp. 471-505. Cambridge University Press, Cambridge - London - New York - Melbourne 1977.

2678 - CRAWFORD, R.M.M., BAINES, M.A.: Tolerance of anoxia and the metabolism of ethanol in tree roots. - New Phytol. 79: 519-526, 1977.

2679 - CRUIZIAT, P.: Untersuchung der Blattwasserkompartimente im Zusammenhang mit der Wiederanfeuchten von Blättern. - In: UNGER, K. (ed.) Biophysikalische Analyse Pflanzlicher Systeme. Pp. 120-130. VEB Gustav Fischer Verlag, Jena 1977.

2680 - CUTLER, J.M., RAINS, D.W.: Effects of irrigation history on responses of cotton to subsequent water stress. - Crop Sci. 17: 329-335, 1977.

2681 - CUTLER, J.M., RAINS, D.W., LOOMIS, R.S.: The importance of cell size in the water relations of plants. - Physiol. Plant. 40: 255-260, 1977.

2682 - CUTLER, J.M., RAINS, D.W., LOOMIS, R.S.: Role of changes in solute concentration in maintaining favorable water balance in field-grown cotton. - Agron. J. 69: 773-779, 1977.

2683 - CZERATZKI, W.: Wasserverbrauch und Trockensubstanzbildung von Mais sowie Bodenevaporation in Abhängigkeit von Saugspannung und Bodenart in Unterdruck-lysimetern im Trockenjahr 1976. - Landbauforsch. Völkenrode 27: 1-14, 1977.

2684 - DA, S., HUBAC, C., VARTANIAN, N.: Influence de la sécheresse sur la morphologie du système racinaire du *Carex setifolia*.- Can. J. Bot. 55: 1236-1245, 1977.

2685 - DĄBROWSKA, J.: Effect of soil moisture on some morphological characters of *Achillea collina* Becker, *A. millefolium* L. ssp. *millefolium* and *A. pannonica* Scheele. - Ekol. Polska 25: 275-288, 1977.

2686 - DADYKIN, V.P., SAMSONOVA, L.P.: O vliyanii plenochnykh antitranspirantov na drevesnye rasteniya. [Effect of film antitranspirants on trees.] - Fiziol. Rast. 24: 574-581, 1977. [In R, ab: E.]

2687 - DALE, R.F., SCHEERINGA, K.L.: The effect of soil moisture on pan evaporation. - Agr. Meteorol. 18: 463-474, 1977.

*2688 - DANCETTE, C.: Cartes d'adaptation a la saison des pluies des mils a cycle court dans la moitie nord du Senegal. - In: Efficiency of Water and Fertilizer Use in Semi-Arid Regions. A Technical Document. Pp. 19-47. International Atomic Energy Agency, Vienna 1976.

2689 - DAS, D.K., JAT, R.L.: Influence of three soil-water regimes on root porosity and growth of four rice varieties. - Agron. J. 69: 197-200, 1977.

2690 - DAS, V.S.R., RAO, I.M., RAGHAVENDRA, A.S.: Mechanism of stomatal movement. - Nature 266: 282, 1977.

2691 - DAS, V.S.R., SANTAKUMARI, M.: Stomatal characteristics of some dicotyledonous plants in relation to the $C_4$ and $C_3$ pathways of photosynthesis. - Plant Cell Physiol. 18: 935-938, 1977.

2692 - DAUTKULOV, A.D., ZHEKSEMBIEVA, R.O.: Vliyanie kaliĭnykh udobreniĭ na nekotorye fiziologicheskie pokazateli u lyutserny. [Effect of potassium fertilizer on some physiological characteristics in alfalfa.] - Fiziol. Biokhim. kul't. Rast. 9: 68-73, 1977. [In R, ab: E.]

2693 - DAVENPORT, D.C., HAGAN, R.M., URIU, K.: Reducing transpiration to conserve water in soil and plants. - California Agr. 31: 40-41, 1977.

2694 - DAVENPORT, T.L.: Auxin involvement in stress induced leaf abscission. - In:
Proc. Fourth Annual' Meeting Plant Growth Regulators Working Group. Pp. 78-79,
Hot Springs 1977.

2695 - DAVENPORT, T.L., JORDAN, W.R., MORGAN, P.W.: Movement and endogenous levels of
abscisic acid during water-stress-induced abscission in cotton seedlings. -
Plant Physiol. 59: 1165-1168, 1977.

2696 - DAVENPORT, T.L., MORGAN, P.W., JORDAN, W.R.: Auxin transport as related to
leaf abscission during water stress in cotton. - Plant Physiol. 59: 554-557,
1977.

2697 - DAVIES, W.J.: Stomatal responses to water stress and light in plants grown
in controlled environments and in the field. - Crop Sci. 17: 735-740, 1977.

2698 - DAVIES, W.J., KOZLOWSKI, T.T.: Variations among woody plants in stomatal
conductance and photosynthesis during and after drought. - Plant Soil 46:
435-444, 1977.

2699 - DAVIS, S.D., VAN BAVEL, C.H.M., McCREE, K.J.: Effect of leaf aging upon stomatal resistance in bean plants. - Crop Sci. 17: 640-645, 1977.

2700 - DAY, W.: A direct reading continuous flow porometer. - Agr. Meteorol. 18: 81
-89, 1977

2701 - DAY, W.: Stomatal resistance in different gases. - J. appl. Ecol. 14: 643-
647, 1977.

2702 - DeBOER, D.W., BROSZ, D.D., WIERSMA, J.L.: Irrigation application depths for
optimum crop production. - Trans. ASAE 20: 1067-1069, 1078, 1977.

*2703 - De BOODT, M., GABRIELS, D.: Soil conservation practices for protection against
erosin in semi-arid regions. - In: Efficiency of Water and Fertilizer Use in
Semi-Arid Regions. A Technical Document. Pp. 49-73. International Atomic
Energy Agency, Vienna 1976.

2704 - De LIS, B.R., CAVAGNARO, J.B.: Modificaciones metabólicas en la planta de
pimiento (*Capsicum annuum* L. cv. Perfection) inducidas por un régimen de se-
quía durante et período crítico de requerimiento hídrico. I - Fracciones
fosforades. [Metabolic changes in pepper plant induced by drought during critical period of water requirement. I - Metabolism of phosphorus.] - Fyton 35:
65-74, 1977. [In Span, ab: E.]

*2705 - DENMEAD, O.T., MILLAR, B.D.: Water transport in wheat. - In: De VRIES, D.A.,
AFGAN, N.H.(ed.): Heat and Mass Transfer in the Biosphere. Part 1. Transfer
Processes in the Plant Environment. Pp. 395-402. Scripta Book Company, Washing-
ton 1975.

*2706 - DERCO, M.: Vplyv závlahy a hnojenia na štruktúru a kvalitu úrody cukrovej re-
py. [The effect of irrigation and fertilization on the structure and quality
of sugar beet produce.] - Rost. Výroba (Praha) 22: 1175-1184, 1976. [In Slov,
ab: E,R.]

2707 - DERCO, M.: Vplyv závlahy a hnojenia na úrodu a ekonomický efekt cukrovej repy.
[Influence of irrigation and fertilization on the yield and economic effective-
ness of sugar beet.] - Rost. Výroba (Praha) 23: 67-76, 1977. [In Slov, ab: E,R.]

2708 - DERCO, M.: Možnosti dalšieho zvyšovania úrody kukurice v podmienkach s regulo-
vaným vodným režimom pódy závlahou. [Possibility of the further increase of
grain maize yield on the basis of moisture regime.] - In: Produkce Biomasy a
Tvorba Výnosu Polních Plodin. Vol. 2. Pp. 65-58. CVTSZ, Praha 1977. [In Slov,
ab: R,E,G.]

*2709 - DHINDSA, R.S., BEWLEY, J.D.: Plant desiccation: polysome loss not due to ribo-
nuclease. - Science 191: 181-182, 1976.

2710 - DHINDSA, R.S., BEWLEY, J.D.: Water stress and protein synthesis. V. Protein
synthesis, protein stability, and menbrane permeability in a drought-sensitive
and a drought-tolerant moss. - Plant Physiol. 59: 295-300, 1977.

2711 - DIANATI, M.: Einfluss von Bewässerung und Plazierung von markierten Stickstoff-
-Düngern auf Ertrag und Stickstoffentzug bei Weizen unter semiariden Klima-
bedigungen im Iran. - Z. Pflanzenernähr. Bodenk. 140: 627-638, 1977.

*2712 - DI FULVIO, T.E.: Observationes en epidermis de *Notocactus* y *Wigginsia* (*Cactaceae*). [Observations of *Notocactus* and *Wigginsia* epidermes.] - Kurtziana 9: 7-17, 1976. [In Span, ab: E.]

2713 - DITTRICH, P., RASCHKE, K.: Malate metabolism in isolated epidermis of *Commelina communis* L. in relation to stomatal functioning. - Planta 134: 77-81, 1977.

2714 - DITTRICH, P., RASCHKE, K.: Uptake and metabolism of carbohydrates by epidermal tissue. - Planta 134: 83-90, 1977.

2715 - DONNELLY, K.J., VANDERLIP, R.L., MURPHY, L.S.: Desiccation of grain sorghum by foliar application of nitrogen solution. - Agron. J. 69: 33-36, 1977.

2716 - DÖRFFLING, K., STREICH, J., KRUSE, W., MUXFELDT, B.: Abscisic acid and after-effect of water stress on stomatal opening potential. - Z. Pflanzenphysiol. 81: 43-56, 1977.

2717 - DORRELL, D.G., CHUBEY, B.B.: Irrigation, fertilizer, harvest dates and storage effects on the reducing sugar and fructose concentrations of Jerusalem artichoke tubers. - Can. J. Plant Sci. 57: 591-596, 1977.

2718 - DOSS, B.D., EVANS, C.E., TURNER, J.L.: Irrigation and applied nitrogen effects on snap bean and pickling cucumbers. - J. Amer. Soc. hort. Sci. 102: 654-657, 1977.

2719 - DOWNTON, W.J.S.: Photosynthesis in salt-stressed grapevines. - Aust. J. Plant Physiol. 4: 183-192, 1977.

2720 - DRAYCOTT, A.P., MESSEM, A.B.: Response by sugar beet to irrigation, 1965-75. - J. agr. Sci. 89: 481-493, 1977.

2721 - DRENNAN, D.S.H., JENNINGS, E.A.: Weed competition in irrigated cotton (*Gossypium barbadense* L.) and groundnut (*Arachis hypogea* L.) in the Sudan Gezira. - Weed Res. 17: 3-9, 1977.

2722 - DREW, M.C., SISWORO, E.J.: Early effects of flooding on nitrogen deficiency and leaf chlorosis in barley. - New Phytol. 79: 567-571, 1977.

2723 - DUBETZ, S.: Effects of high rates of nitrogen on Neepawa wheat grown under irrigation. I. Yield and protein content. - Can. J. Plant Sci. 57: 331-336, 1977.

*2724 - DUC, T.M.: Irrigation de la zone centre - nord du Senegal - resultats de recherches et perspectives. - In: Efficiency of Water and Fertilizer Use in Semi-Arid Regions. A Technical Document. Pp. 249-271. International Atomic Energy Agency, Vienna 1976.

2725 - DUMBROFF, E.B., BREWER, W.L.J.: Mechanisms of adjustement in tomato following release from osmotic stress. - Z. Pflanzenphysiol. 81: 167-172, 1977.

2726 - DUMBROFF, E.B., BROWN, D.C.W., THOMPSON, J.E.: Effect of senescence on levels of free abscisic acid and water potentials in cotyledons of bean. - Bot. Gaz. 138: 261-265, 1977.

*2727 - DUNHAM, R.J., NYE, P.H.: The influence of soil water content on the uptake of ions by roots. III. Phosphate, potassium, calcium and magnesium uptake and concentration gradients in soil. - J. appl. Ecol. 13: 967-984, 1976.

2728 - DUNIWAY, J.M.: Changes in resistance to water transport in safflower during the development of *Phytophthora* root rot. - Phytopathology 67: 331-337, 1977.

*2729 - DVORAKOVSKIĬ, M.S., NUKHIMOVSKAYA, Yu.D.: Povedenie podrosta pikhty sibirskoĭ i eli obyknovennoĭ pri ikh sovmestnom proizrastanii v Moskovskoĭ oblasti. [Behaviour of undergrowth of fir and spruce under combined cultivation in region of Moscow.] - Nauch. Dokl. vyssh. Shkoly, biol. Nauki 1976 (4): 90-96, 1976. [In R.]

2730 - EDER, A., HUBER, W.: Zur Wirkung von Abscisinsäure und Kinetin auf biochemische Veränderungen in *Pennisetum typhoides* unter Stresseinwirkungen. - Z. Pflanzenphysiol. 84: 303-311, 1977.

2731 - EDWARDS, M., MEIDNER, H.: Direct measurements of turgor pressure potentials. III. (A) injections of solutions into stomatal cells, and (B) comparisons of pressure potentials in stomatal cells of different species. - J. exp. Bot. 28: 669-677, 1977.

*2732 - EENINK, A.H., ALVAREZ, J.M.: Indirect selection for tetraploidy in lettuce (*Lactuca sativa* L.). - Euphytica 24: 661-668, 1975.

2733 - EGOROV, V.G., LEBEDEV, G.V., DUMBADZE, V.Z., BRYUKVIN, V.G.: Pogloshchenie vody rasteniyami v usloviyakh sukhoveya i pri periodicheskom opryskivanii ikh vodoĭ. [Absorption of water by plants under conditions of arid wind and during periodic spraying them with water.] - Fiziol. Rast. 24: 188-190, 1977. [In R.]

2734 - ELIÁŠ, P.: Ekofyziologické štúdium difúznych odporov listov rastlín dubo--hrabového lesa na výskumnej ploche IBP v Bábe pri Nitre. [Ecophysiological studies of the leaf transpiration resistances of plants of oak-hornbeam forest on IBP research area at Báb near Nitra.] - In: Dni Rastlinnej Fyziológie. I. Pp. 172-175, 309. Slovenská Botanická Spoločnost SAV, Bratislava 1977. [In Slov, ab: R,E.]

2735 - ELIÁŠ, P.: Aktivita prieduchov duba zimného v prirodzenom prostredí. [Stomata activity of *Quercus petraea* in natural environment.] - In: HUZULÁK , J., MASAROVIČOVÁ, E.(ed.): Fotosyntéza a Vodný Režim Drevín. Pp. 84-91. Modra--Piesky 1977. [In Slov, ab: E,R.]

2736 - ELIÁŠ, P.: Možnosti zniženia transpirácie chemickými látkami. [Possibilities of reduction of transpiration by chemical compounds.] - In: REPKA, J. (ed.): Zborník Referátov zo Seminára Fyziologicko-Genetické a Chemické Faktory Produktivity Rastlín. Pp. 117-126. Vysoká Škola Polnohospodárská, Nitra 1977. [In Slov.]

2737 - ELIÁŠ, P.: Vodivost prieduchov javora poľného v lesných podmienkách. [Stomatal conductance of *Acer campestre* L. under forest conditions.] - Acta Musei Silesiae, Ser. dendrol. 26: 9-37, 1977. [In Slov, ab: E.]

*2738 - EL-KEREDY, M.S., EL-SHOUNY, K.A.: Breeding for saline-resistant varieties of rice. Comparative performance of some upland and paddy rice varieties. - Z. Pflanzenzücht. 76: 231-239, 1976.

2739 - EL-SHARKAWI, H.M., MICHEL, B.E.: Effects of soil water matric potential and air humidity on $CO_2$ and water vapor exchange in two grasses. - Photosynthetica 11: 176-182, 1977.

2740 - EL-SHARKAWI, H.M., SALAMA, F.M.: Effects of drought and salinity on some growth-contributing parameters in wheat and barley. - Plant Soil 46: 423-433, 1977.

*2741 - EL-SHARKAWY, M.A.: A note on methods measuring the internal water status in leaves of sunflower plants (*Helianthus annuus* L.). - Libyan J. Agr. 4: 43-46, 1975.

2742 - ELSTON, J., DENNETT, M.D.: A weather watch for semi-arid lands within the tropics. - Phil. Trans. roy. Soc. London B 278: 593-609, 1977.

2743 - EMMINGHAM, W.H., WARING, R.H.: An index of photosynthesis for comparing forest sites in western Oregon. - Can. J. Forest Res. 7: 165-174, 1977.

*2744 - ENDO, J., NAKAMURA, T., NAGASAWA, M.: Geographical variation of *Asarum caulescens* maxim. - J. Jap. Bot. 51: 209-219, 1976. [In Jap., ab: E.]

*2745 - ENDO, M.: [Studies on the daily change of fruit size of the Japanese pear. VI. Diurnal fluctuation of fruit diameter as affected by soil moisture.] - J. Jap. Soc. hort. Sci. 44: 248-259, 1975. [In Jap, ab: E.]

*2746 - ENDO, M., OGASAWARA, S.: [Studies on the daily change of fruit size of the Japanese pear. V. Diurnal fluctuation of fruit size as affected by rainfall or water sprinkling.] - J. Jap. Soc. hort. Sci. 43: 359-367, 1975. [In Jap, ab: E.]

*2747 - ENIKEEV, S.G.: Nekotorye fiziologicheskie pokazateli sakharnoĭ svekly pri oroshenii v Tatarii i ee produktivnost'. [Some physiological characteristics of irrigated sugar beet in Tartaria and its productivity.] - Tr. Gor'kov. sel'skokhoz. Inst. 69 (Botanika i Fiziologiya Rasteniĭ): 138-142, 1976. [In R.]

*2748 - ENIKEEV, S.G., MAZIL'NIKOV, G.V., USHAKOV, V.Yu., KIYAMOV, A.G.: Izmenenie fo-
tosinteticheskoĭ aktivnosti list'ev gorokha pod deĭstviem poverkhnostno-aktiv-
nykh veshchestv. [Changes in photosynthetic activity of pea leaves affected
by surface-active substances.] - Tr. Gor'kov. sel'skokhoz. Inst. 69 (Botanika
i Fiziologiya Rasteniĭ): 84-86, 1976. [In R.]

*2749 - ENIKEEV, S.G., MAZIL'NIKOV, G.V., USHAKOV, V.Yu., ZALYAEV, Z.K.: Deĭstvie po-
vyshennykh temperatur i ouabaina na raspredelenie kaliya i fotosintez v kletkakh list'ev gorokha. [Effect of elevated temperatures and ouabain on potassium
distribution and photosynthesis in cells of pea leaves.] - Tr. Gor'kov. sel'-
skokhoz. Inst. 69 (Botanika i Fiziologiya Rasteniĭ): 81-83, 1976. [In R.]

*2750 - ENIKEEV, S.G., USHAKOV, V.Yu., GARAFEEV, A.G.: Vzaimosvyaz fotosinteza i lipi-
doobrazovaniya v list'yakh gorokha pri vodnom defitsite. [Relationship between
rate of photosynthesis and formation of lipids in pea leaves under water stress.]
- Tr. Gor'kov. sel'skokhoz. Inst. 69 (Botanika i Fiziologiya Rasteniĭ): 91-96,
1976. [In R.]

2751 - ENOCH, H.Z., HURD, R.G.: Effect of light intensity, carbon dioxide concentration, and leaf temperature on gas exchange of spray carnation plants. - J. exp.
Bot. 28: 84-95, 1977.

2752 - EPSTEIN, E., NORLYN, J.D.: Seawater-based crop production: a feasibility study.
- Science 197: 249-251, 1977.

2753 - EVANS, L.S., GMUR, N.F., Da COSTA, F.: Leaf surface and histological perturba-
tions of leaves of *Phaseolus vulgaris* and *Helianthus annuus* after exposure to
simulated acid rain. - Amer. J. Bot. 64: 903-913, 1977.

2754 - EVANS, P.S., URIU, K., PEARSON, J.R.: Some effects of potassium deficiency on
water relations of French prune. - J. Amer. Soc. hort. Sci. 102: 648-650, 1977.

2755 - EVERT, R.F.: Phloem structure and histochemistry. - Annu. Rev. Plant Physiol.
28: 199-222, 1977.

2756 - EZE, J.M.O., BERRIE, G.K.: Further investigations into the physiological relationship between an epiphyllous liverwort and its host leaves. - Ann. Bot. 41:
351-358, 1977.

2757 - FAIZ, S.M.A., WEATHERLEY, P.E.: The location of the resistance to water movement in the soil supplying the roots of transpiring plants. - New Phytol. 78:
337-347, 1977.

2758 - FALKE, H., GÖRLITZ, H., BERGMANN, W.: Der Einfluss der Beregnung auf die Mi-
kronährstoffdynamik der Böden und den Mikronährstoffgehalt der Pflanzen. -
Arch. Acker- Pflanzenbau Bodenk. 21: 715-723, 1977.

2759 - FEDAK, G., MACK, A.R.: Influence of soil moisture levels and planting dates on
yield and chemical fractions in two barley cultivars. - Can. J. Plant Sci. 57:
261-267, 1977.

*2760 - FEDDES, R A , Van WIJK, A.L.M.: An integrated model-approach to the effect of
water management on crop yield. - Agr. Water Manage. 1: 3-20, 1976.

2761 - FEDERER, C.A.: Leaf resistance and xylem potential differ among broadleaved
species. - Forest Sci. 23: 411-418, 1977.

2762 - FEDOROVA, A.I.: Vliyanie neblagopriyatnykh usloviĭ sredy na uroven' abtsisovoĭ
kisloty u listvennitsy sibirskoĭ. [The effect of unfavourable environmental
conditions on the level of abscisic acid in Siberian larch.] - Fiziol. Rast.
24: 1223-1227, 1977. [In R, ab: E.]

2763 - FEDOROVSKAYA, M.D., SHTRAUSBERG, D.V., VAKHMISTROV, D.B., SOLOV'EV, V.A.:
Svet i temnota kak datchiki vremeni tsirkadnogo ritma placha podsolnechnika.
[Light and darkness as timers of circade rhythm of sunflower exudation.] -
Fiziol. Rast. 24: 83-89, 1977. [In R, ab: E.]

2764 - FENSOM, D.S.: A role of electrogenic pumps in producing turgor in *Nitella*. -
In: MARRÉ, E., CIFERRI, O. (ed.): Regulation of Cell Membrane Activities in
Plants. Pp. 91-102. Elsevier/North Holland Biomedical Press, Amsterdam - Oxford
- New York 1977.

2765 - FENSOM, D.S., ROSS, S.M.: Note on lack of effect of ethylene on water permeability, electroosmotic efficiency, or transcellular water flow of the plasma membranes in *Nitella*. - Can. J. Bot. 55: 615-616, 1977.

2766 - FENTON, R., DAVIES, W.J., MANSFIELD, T.A.: The role of farnesol as a regulator of stomatal opening in *Sorghum*. - J. exp. Bot. 28: 1043-1053, 1977.

2767 - FERHI, A., LETOLLE, R,: Variation de la composition isotopique de l'oxygène organique de quelques plantes en fonction de leur milieu de vie. - C.R. Acad. Sci. Paris, Sér. D 284: 1887-1889, 1977.

2768 - FERHI, A., LETOLLE, R.: Transpiration and evaporation as principal factors in oxygen isotope variations of organic matter in land plants. - Physiol. vég. 15: 363-370, 1977.

2769 - FERNANDEZ, F.G., CARO, M., CERDÁ, A.: Influnce of NaCl in the irrigation water on yield and quality of sweet pepper (*Capsicum annuum*). - Plant Soil 46: 405-413, 1977.

*2770 - FERREIRO, E.A., HELMY, A.K.: Positive and negative root membrane potentials. - An. Edafol. Agrobiol. 35: 95-102, 1976.

2771 - FERRIER, J.M., DAINTY, J.: Water flow in *Beta vulgaris* storage tissues. - Plant Physiol. 60: 662-665, 1977.

2772 - FERRIER, J.M., DAINTY, J.: A new method for measurement of hydraulic conductivity and elastic coefficients in higher plant cells using an external force. - Can. J. Bot. 55: 858-866, 1977.

2773 - FERRON, F., COSTES, C.: Study of the efficiency of Hydrasyl as an antitranspirant with an isotopic method. - Physiol. Plant. 39: 196-200, 1977.

2774 - FFOLLIOTT, F., THORUD, D.B.: Water yield improvement by vegetation management. - Water Resour. Bull. 13: 563-571, 1977.

2775 - FINCK, A.: Soil salinity and plant nutritional status. - In: DREGNE, H.E. (ed.): Managing Saline Water for Irrigation. Pp. 199-210. Texas Technical University, Lubbock 1977.

2776 - FISCHER, R., DENGLER, N.G.: Mesophyll cell walls in hemlock, *Tsuga canadensis*. - Can. J. Bot. 55: 1510-1515, 1977.

2777 - FISCHER, R.A., LINDT, J.H., GLAVE, A.: Irrigation of dwarf wheats in the Vaqui Valley of Mexico. - Exp. Agr. 13: 353-367, 1977.

2778 - FISCHER, R.A., SANCHEZ, M., SYME, J.R.: Pressure chamber and air flow porometer for rapid field indication of water status and stomatal condition in wheat. - Exp. Agr. 13: 341-351, 1977.

2779 - FISCUS, E.L.: Determination of hydraulic and osmotic properties of soybean root systems. - Plant Physiol. 59: 1013-1020, 1977.

2780 - FISCUS, E.L.: Effects of coupled solute and water flow in plant roots with special reference to Brouwer's experiments. - J. exp. Bot. 28: 71-77, 1977.

2781 - FITZGERALD, P.D., KNIGHT, T.L., JANSON, C.G.: Lucerne irrigation on light soils. - N. Zeal. J. exp. Agr. 5: 23-27, 1977.

2782 - FLINN, A.M., ATKINS, C.A., PATE, J.S.: Significance of photosynthetic and respiratory exchanges in the carbon economy of the developing pea fruit. - Plant Physiol. 60: 412-418, 1977.

2783 - FLORES, E.M., ESPINOZA, A.M.: Ultrastructura foliar de *Vigna unguiculata* L. - [Ultrastructure of the leaf of *Vigna unguiculata*.] - Rev. Biol. trop. 25: 159-169, 1977. [In Span, ab: E.]

2784 - FLORES, E.M., ESPINOZA, A.M., KOZUKA, Y.: Observationes sobre la epidermis foliar de *Zea mays* L. al microscopio electrónico de rastrero. [Scanning electron microscope observations of the leaf epidermis of *Zea mays*.] - Rev. Biol. trop. 25: 123-135, 1977. [In Span, ab: E.]

2785 - FLORES, E.M., ESPINOZA, A.M., KOZUKA, Y.: Estudio ultrastructural de la epidermis foliar de *Phaseolus vulgaris* L. [Structural study of the foliar epidermis of *Phaseolus vulgaris*.] - Turrialba 27: 117-124, 1977. [In Span, ab: E.]

2786 - FLOWERS, T.J., TROKE, P.F., YEO, A.R.: The mechanism of salt tolerance in ha-
       lophytes. - Annu. Rev. Plant Physiol. 28: 89-121, 1977.

2787 - FLÜCKIGER, W., FLÜCKIGER-KELLER, H., OERTLI, J.J., GUGGENHEIM, R.: Verschmutzung
       von Blatt- und Nadeloberflächen im Nahbereich einer Autobahn und deren Einfluss
       auf den stomatären Diffusionswiderstand. - Europ. J. Forest Pathol. 7: 358-
       364, 1977.

2788 - FLÜHLER, H., ARDAKANI, M.S., SZUSZKIEWICZ, T.E., STOLZY, L.H.: Field-measured
       water uptake of sudangrass roots as affected by fertilization. - Agron. J. 69:
       269-274, 1977.

2789 - FORDE, B.J., MITCHELL, K.J., EDGE, E.A.: Effect of temperature, vapour-pressu-
       re deficit and irradiance on transpiration rates of maize, paspalum, Westerwolds
       and perennial ryegrass, peas, white clover and lucerne. - Aust. J. Plant Phy-
       siol. 4: 889-899, 1977.

2790 - FOULDS, W.: The physiological response to moisture supply of cyanogenic and
       acyanogenic phenotypes of *Trifolium repens* L. and *Lotus corniculatus* L. -
       Heredity 39: 219-234, 1977.

2791 - FOULDS, W., YOUNG, L.: Effect of frosting, moisture stress and potassium cya-
       nide on the metabolism of cyanogenic and acyanogenic phenotypes of *Lotus cor-
       niculatus* L. and *Trifolium repens* L. - Heredity 38: 19-24, 1977.

2792 - FRANGMEIER, D.D., MOHAMMED, R.A.: Irrigation management of short-season, hight
       -population cotton. - Trans. ASAE 20: 869-872, 1977.

2793 - FRANICH, R.A., WELLS, L.G., BARNETT, J.R.: Variation with tree age of needle
       cuticle topography and stomatal structure in *Pinus radiata* D. Don. - Ann. Bot.
       41: 621-626, 1977.

2794 - FRANK, A.B., HARRIS, D.G., WILLIS, W.O.: Growth and yield of spring wheat as
       influenced by shelter and soil water. - Agron. J. 69: 903-906, 1977.

2795 - FRANK, A.B., HARRIS, D.G., WILLIS, W.O.: Plant water relationships of spring
       wheat as influenced by shelter and soil water. - Agron. J. 69: 906-910, 1977.

2796 - FRANQUIN, P., FOREST, F.: Des programmes pour l'evaluation et l'analyse fre-
       quentielle des termes du bilan hydrique. - Agron. trop. 32: 7-11, 1977.

*2797 - FREEMAN, B.M., BLACKWELL, J., GARZOLI, K.V.: Irrigation frequency and total
       water application with trickle and furrow systems. - Agr. Water Manage. 1:
       21-31, 1976.

2798 - FRIMMEL, G.: Der Einfluss des Mehltaubefalles auf Winterfestigkeit und Dürre-
       resistenz. - Z. Pflanzenzücht. 79: 256-260, 1977.

2799 - FRITSCHEN, L.J., HSIA, J., DORAISWAMY, P.: Evapotranspiration of a Douglas fir
       determined with a weighing lysimeter. - Water Resour. Res. 13: 145-148, 1977.

2800 - FRÖLICH, W.G., POLLMER, W.G., KLEIN, D.: Performance of isogenic liquleless-2
       maize hybrids as influenced by irrigation, row width, plant density and nitro-
       gen fertilizer. - Z. Acker- Pflanzenbau 145: 207-223, 1977.

*2801 - FROMMHOLD, I.: Physiologie der Gramineen-Stomata. I. Struktur, Funktion und
       Regulation der Öffnungsweite. - Wiss. Z. Humboldt-Univ. Berlin, math.- natur-
       wiss. Reihe 25: 803-810, 1976.

*2802 - FROMMHOLD, I.: Physiologie der Gramineen-Stomata. II. Beziehungen zwischen der
       Stomatafrequenz und -apertur sowie der Transpirations- bzw. Photosyntheserate
       - Wiss. Z. Humboldt-Univ. Berlin, math.-naturwiss. Reihe 25: 811-817, 1976.

*2803 - FUCHS, M., STANHILL, G., MORESHET, S.: Effect of increasing foliage and soil
       reflectivity on the solar radiation balance of wide-row grain sorghum. - Agron.
       J. 68: 865-871, 1976.

2804 - GAFF, D.F.: Desiccation tolerant vascular plants of southern Africa. - Oecolo-
       gia 31: 95-109, 1977.

 805 - GALATIS, B.: Differentiation of stomatal meristomoids and guard cell mother
       cells into guard-like cells in *Vigna sinensis* leaves after colchicine treat-
       ment. An ultrastructural and experimental approach. - Planta 136: 103-114,
       1977.

2806 - GALLAHER, R.N.: Soil water use and yield of corn and soybeans no-till planted
       in rye. - Georgia agr. Res. 19: 14-17, 1977.

2807 - GALLAHER, R.N.: Soil moisture conservation and yield of crops no-till planted
       in rye. - Soil. Sci. Soc. Amer. J. 41: 145-147, 1977.

*2808 - GANCHARYK, M.M., LYAGENCHANKA, B.I., MIKUL'SKAYA, S.A., TALANAVA, K.S.,
       MATSYUSHEISKAYA, V.P.: Uplyŭ impul'snaga dazhdzhavannya na fiziyalogiyu i
       praduktsyinasts' aŭsa pry roznai vil'gotnastsi tarfyanoĭ gleby. [The effect
       of impulse irrigation on physiology and productivity of oat under different
       moisture of peat soil.] - Vestsi Akad. Navuk Belarus. SSR, Ser. biyal. Navuk
       1976 (6): 5-11, 137, 1976. [In Belorus, ab: R.]

2809 - GARBER, M.P.: Effect of light and chilling temperatures on chilling-sensitive
       and chilling-resistant plants. Pretreatment of cucumber and spinach thylakoids
       in vivo and in vitro. - Plant Physiol. 59: 981-985, 1977.

*2810 - GARDNER, W.R., JURY, W.A., KNIGHT, J.: Water uptake by vegetation. - In:
       De VRIES, A.D., AFGAN, N.H. (ed.): Heat and Mass Transfer in the Biosphere.
       Part 1. Transfer Processes in the Plant Environment. Pp. 443-456. Scripta
       Book Company, Washington 1975.

2811 - GARSED, S.G., READ, D.J.: Sulphur dioxide metabolism in soy-bean, *Glycine max.*
       var. biloxi. I. The effect of light and dark on the uptake and translocation
       of $^{35}SO_2$. - New Phytol. 78: 111-119, 1977.

2812 - GÄRTNER, M.: Biologische Grundlagen zu einer Modellierung der Stoffproduktion
       und Energienutzung in einem Wasserpflanzenbestand von *Typha latifolia* and
       *Typha angustifolia.* - In: UNGER, K. (ed.): Biophysikalishe Analyse Pflanzlicher
       Systeme. Pp. 269-282. VEB Gustav Fischer Verlag, Jena 1977.

*2813 - GAUL, B.L., ABROL, I.P., DARGAN, K.S.: Note on the irrigation needs of *Ses-
       bania aculeata* Poir. for greenmanuring during summer. - Ind. J. agr. Sci. 46:
       434-436, 1976.

2814 - GAUSMAN, H.W.: Reflectance of leaf components. - Remote Sensing Environ. 6:
       1-9, 1977.

2815 - GAUSMAN, H.W., LEAMER, R.W., NORIEGA, J.R., RODRIGUEZ, R.R., WIEGAND, C.L.:
       Field-measured spectrophotometric reflectances of disked and nondisked soil
       with and without wheat straw. - Soil Sci. Soc. Amer. J. 41: 793-796, 1977.

2816 - GAY, A.P., LOACH, K.: Leaf conductance changes on leafy cuttings of *Cornus*
       and *Rhododendron* during propagation. - J. hort. Sci. 52: 509-516, 1977.

2817 - GAY, A.P., NICHOLS, R.: The effects of some chemical treatments on leaf water
       conductance of cut, flowering stems of *Chrysanthemum morifolium.* - Scientia
       Hort. 6: 167-177, 1977.

2818 - GEISLER, G., AFSCHAR, I.: Der Einfluss der Wasserversorgung und Stickstoff-
       düngung auf die Entwicklung der generativen Organe bei Mais. - Z. Acker-
       Pflanzenbau 145: 257-271, 1977.

*2819 - GEJ, B.: Charakterystyka fizjologiczna poszczególnych liści salaty traktowa-
       nej symazyną. [Physiological characteristics of individual leaves of lettuce
       treated with simazine.] - Roczn. Nauk. roln. A 101 (4): 7-19, 1976. [In Pol,
       ab: E, R.]

2820 - GENKEL', P.A., SATAROVA, N.A., SHAPOSHNIKOVA, S.V.: Sintez belka u poĭkilo-
       kserofitov i zarodysheĭ pshenitsy v nachal'nyĭ period nabukhaniya. [Protein
       synthesis in poikiloxerophytes and wheat embryos at the beginning of swelling.]
       - Fiziol. Rast. 24: 906-912, 1977. [In R, ab: E.]

*2821 - GEORGIEV, M.: The influence of the soil moisture on the content and distribu-
       tion of B, Cu, Mg, Fe and Mn in the above-ground parts of winter wheat. - Acta
       bot. Croat. 34: 180, 1975.

*2822 - GEORGIEV, M., SPASENOSKI, M.: Influence of the soil moisture on the content
       of phosphorus in the leaves, stalks and peels phosphorus fractions in the grain
       and the chlorophyll pigment in the leaves of winter wheat.. - Acta bot. Croat.
       34: 181, 1975.

2823 - **GHOSH, R.K.**: Determination of unsaturated hydraulic conductivity from moisture retention function. - Soil Sci. 124: 122-124, 1977.

2824 - **GICHNER, T., VELEMÍNSKÝ, J., POKORNÝ, V.**: Changes in the yield of genetic effects and DNA single-strand breaks during storage of barley seeds after treatment with diethyl sulphate. - Environ. exp. Bot. 17: 63-67, 1977.

2825 - **GILISSEN, L.J.W.**: Style-controlled wilting of the flower. - Planta 133: 275-280, 1977.

2826 - **GILISSEN, L.J.W.**: The influence of relative humidity on the swelling of pollen grains in vitro. - Planta 137: 299-301, 1977.

2827 - **GILLES, R., PEQUEUX, A.**: Effect of salinity on the free amino acids pool of the red alga *Porphyridium purpureum (= P. cruentum)*. - Comp. Biochem. Physiol. 57: 183-185, 1977.

2828 - **GILLHAM, F.E.M., HARRIGAN, E.K.S.**: Disease resistant flue-cured tobacco breeding lines for north Queensland. 2. Resistance to bacterial wilt, *Pseudomonas solanacearum* and black shank, *Phytophthora nicotianae* var. nicotianae. - Aust. J. exp. Agr. anim. Husb. 17: 659-663, 1977.

2829 - **GILMORE, A.R.**: Effects of soil moisture stress on monoterpens in loblolly pine. - J. Chem. Ecol. 3: 667-676, 1977.

2830 - **GINDRAT, D.**: Effets de concentrations élevées de sels sur la croissance, la sporulation et la pigmentation de *Trichoderma* spp. - Can. J. Microbiol. 23: 607-616, 1977.

2831 - **GINZO, H.D., CARCELLER, M.S., FONSECA, E.**: CCC [(2-chloroethyl)trimethyl-ammonium chloride] and the regulation of plant water status in wheat (*Triticum aestivum* L.). - Fyton 35: 85-92, 1977.

2832 - **GISI, U., OERTLI, J.J., SCHWINN, F.J.**: Wasser- und Salzbeziehungen der Sporangien von *Phytophthora cactorum* (Leb. et Cohn) Schroet. in vitro. - Phytopathol. Z.: 89: 261-284, 1977.

*2833 - **GIVNISH, T.J., VERMEIJ, G.J.**: Sizes and shapes of liane leaves. - Amer. Natur. 110: 743-778, 1976.

2834 - **GLINKA, Z.**: Effects of abscisic acid and of hydrostatic pressure gradient on water movement throught excised sunflower roots. - Plant Physiol. 59: 933-935, 1977.

2835 - **GLOSER, J.**: Photosynthesis and respiration of some alluvial meadow grasses: responses to soil water stress, diurnal and seasonal courses. - Přírodověd. Práce Ústavů Českoslov. Akad. Věd Brně 11(4): 1-34, 1977.

2836 - **GOH, C.J., AVADHANI, P.N., LOH, C.S., HANEGRAAF, C., ARDITTI, J.**: Diurnal stomatal and acidity rhythms in orchid leaves. - New Phytol. 78: 365-372, 1977.

2837 - **GOLCZ, L., ZAŁECKI, R.**: Wpływ wzrastających dawek azotu przy różnym uwilgotnieniu i odczynie gleby na ilość i jakość plonu sporyszu (*Claviceps purpurea* Tul.). [Effect of increasing doses of nitrogen in soil of various moisture content and pH value on the quantity and quality of ergot crop.] - Herba Polon. 23: 143-148, 1977. [In Pol, ab: E, R.]

2838 - **GOLOVATYĬ, V.G., KHUDYAKOVA, Kh.K.**: Optimizatsiya vodnogo rezhima i mineral'-nogo pitaniya dlya ezhi sbornoĭ. [Optimization of water regime and mineral nutrition of dewgrass.] - Fiziol. Rast. 24: 773-778, 1977. [In R, ab: E.]

2839 - **GORA, A., SCHWARZ, K., SEIFERT, U.**: Pflanzenbedarfsgerechte Bodenwasserregulierung durch unterirdische wechselseitige Wasserregulierung (UWWR) im Vergleich zu anderen Verfahren der Ent- und Bewässerung auf einem Schwarzstaugley - Standort. - Arch. Acker- Pflanzenbau Bodenk. 21: 87-98, 1977.

2840 - **GOSSE, G., PERRIER, A., ITIER, B.**: Etude de l'évapotranspiration réelle d'une culture de blé dans le bassin parisien. - Ann. agron. 28: 521-541, 1977.

2841 - **GRABHERR, G.**: Der $CO_2$-Gaswechsel des immergrünen Zwergstrauches *Loiseleuria procumbens* (L.) Desv. in Abhängigkeit von Strahlung, Temperatur, Wasserstress und phänologischen Zustand. - Photosynthetica 11: 302-310, 1977.

2842 - GRABHERR, G., CERNUSCA, A.: Influence of radiation, wind, and temperature on $CO_2$ gas exchange of the alpine dwarf shrub community *Loiseleurietum cetrariosum*. - Photosynthetica 11: 22-28, 1977.

2843 - GRACE, J.: Plant Response to Wind. (Experimental Botany: An International Series of Monographs. Vol. 13). Academic Press, London - New York - San Francisco 1977.

2844 - GRACE, J., RUSSELL, G.: The effect of wind on grasses III. Influence of continuous drought or wind on anatomy and water relations in *Festuca arundinacea* Schreb. - J. exp. Bot. 28: 268-278, 1977.

2845 - GRADMANN, D.: Potassium and turgor pressure in plants. - J. theor. Biol. 65: 597-599, 1977.

2846 - GRANIN, A.V., PRONINA, N.D., VESELOVSKIĬ, V.A.: Vliyanie obezvozhivaniya i peregreva na poslesvechenie fotosinteticheskogo apparata poĭkilogidrovykh i gomeogidrovykh rasteniĭ. [The effect of dehydration and superheating on delayed light emission by the photosynthetic apparatus of poikilohydrous and homeohydrous plants.] - Fiziol. Rast. 24: 1261-1268, 1977. [In R, ab: E.]

2847 - GRANT, N.G., WALSBY, A.E.: The contribution of photosynthate to turgor pressure rise in the planktonic blue-green alga *Anabaena flos-aquae*. J. exp. Bot. 28: 409-415, 1977.

2848 - GREBANIER, A.E., JAGENDORF, A.T.: Irreversible uncoupling of spinach chloroplasts by sulfate and ADP. - Plant Cell Physiol. 1977 (Spec. Issue 3 - Photosynthetic Organelles. Structure and Function): 103-114, 1977.

2849 - GREEN, M.S., ETHERINGTON, J.R.: Oxidation of ferrous iron by rice (*Oryza sativa* L.) roots: a mechanism for waterlogging tolerance? - J. exp. Bot. 28: 678-690, 1977.

2850 - GREENE, D.W., BUKOVAC, M.J.: Foliar penetration of naphthaleneacetic acid: enhancement by light and role of stomata. - Amer. J. Bot. 64: 96-101, 1977.

2851 - GREGORY, S.C.: A simple technique for measuring the permeability of coniferous wood and its application to the study of water conduction in living trees. - Europ. J. Forest Pathol. 7: 321-328, 1977.

2852 - GREGORY, S.C.: The effect of *Peridermium pini* (Pers.) Lev. on water conduction in *Pinus sylvestris* L. - Europ. J. Forest Pathol. 7: 328-338, 1977.

2853 - GREILICH, J., WENDLING, U.: Ergebnisse der Eichungen von Neutronen-Bodenfeuchtenson den unterschiedlicher Messanordung (Geometrie) in Behältern unter Berücksichtigung des Bodenprofils von Löss-Schwarzerde. - Arch. Acker- Pflanzenbau Bodenk. 21: 319-325, 1977.

2854 - GRIER, C.C., RUNNING, S.W.: Leaf area of mature northwestern coniferous forests: Relation to site water balance. - Ecology 58: 893-899, 1977.

2855 - GRIFFIN, D.M.: Water potential and wood-decay fungi. - Annu. Rev. Phytopathol. 15: 319-329, 1977.

2856 - GRIMES, D.W., DICKENS, W.L.: Cotton responses to irrigation. - California Agr. 31: 16-17, 1977.

2857 - GRINEVA, G.M., NECHIPORENKO, G.A.: Raspredelenie i prevrashchenie sakharozy-
-U-$C^{14}$ u rastenii kukuruzy v usloviyakh zatopleniya. [Distribution and transformation of sucrose-U-$C^{14}$ in maize plants in conditions of inundation.] - Fiziol. Rast. 24: 44-50, 1977. [In R, ab: E.]

2858 - GROUT, B.W.W., ASTON, M.J.: Transplanting of cauliflower plants regenerated from meristem culture. 1. Water loss and water transfer related to changes in leaf wax and to xylem regeneration. - Hort. Res. 17: 1-7, 1977.

2859 - GUBBELS, G.H.: Interaction of cultivar, sowing date and sowing rate of lodging, yield, and seed weight of buckwheat. - Can. J. Plant Sci. 57 : 317 - 321, 1977.

*2860 - GÜCKEL, W., SYNNATSCHKE, G.: Techniques for measuring the wetting of leaf surfaces. - Pestic. Sci. 6: 595-603, 1975.

*2861 - GULYAEV, B.I.: Vliyanie abstsizovoĭ kisloty na ust'ichnye dvizheniya i fotosintez. [Effect of abscisic acid on stomatal movements and photosynthesis.] - Dokl. Akad. Nauk. Ukr. SSR, Ser.B 11: 1022-1026, 1976. [In R, ab. L.]

2862 - GULYAEV, B.I.: Relaksatsionnye avtokolebaniya gazoobmena list'ev. [Self-excited relaxation oscillation of gas exchange in leaves.] - Fiziol. Biokhim. kul't. Rast. 9: 520-526, 1977. [In R, ab: E.]

*2863 - GULYAEV, B.I., TKACHUK, K.S.: Diya i pislyadiya gruntovoĭ posukhi na fotosintez i difuziĭni opori listkiv ozimoĭ pshenitsy. [Effect and after-effect of soil drought on photosynthesis and diffusion resistance of winter wheat leaves.] - Dopov. Akad. Nauk Ukr. RSR, Ser. B 12: 1111-1114, 1976. [In Ukr, ab: E.]

*2864 - GULYAEV, O.S.: O probleme melioratsii klimata v bogarnom zemledelii yuga Zapadnoĭ Sibiri i Severnogo Kazakhstana. [On the problem of climate reclamation in dry farming in the south of the Western Siberia and North Kazakhstan.] - Izv. Sib. Otd. Akad. Nauk SSSR, Ser. biol. Nauk 1976 (10): 3-10, 1976. [In R, ab: E.]

2865 - GUPTA, R.K.: A note on photosynthesis in relation to water content in liverworts: *Porella platyphylla* and *Scapania undulata*. - Aust. J. Bot. 25: 363-365, 1977.

2866 - GUPTA, R.K.: A study of photosynthesis and leakage of solutes in relation to the desiccation effects in bryophytes. - Can. J. Bot. 55: 1186-1194, 1977.

2867 - GUPTA, R.K.: An artefact in studies of the responses of respiration of bryophytes to desiccation. - Can. J. Bot. 55: 1195-1200, 1977.

*2868 - GUPTA, S.C., RANI, S., RAJESWARI, V.M.: Stomata on foliar and floral parts of *Alangium salviifolium*. - Phytomorphology 26: 390-395, 1976.

2869 - GUTCHNECHT, J., BISSON, M.A.: Ion transport and osmotic regulation in giant algal cells. - In: JUNGREIS, A.M., HODGES, T.K., KLEINZELLER, A., SCHULTZ, S.G. (ed.): Water Relations in Membrane Transport in Plants and Animals. Pp. 3-14, Academic Press, New York - San Francisco - London 1977.

*2870 - GYMER, P.T., WHITTINGTON, W.J.: Factors influencing the proportion of natural hybrids between *Lolium perenne* L. and *Festuca pratensis* Huds. in permanent pastures. - J. Brit. Grassland Soc. 31: 165-169, 1976.

2871 - HACHUM, A.Y., ALFARO, J.F.: Water infiltration and runoff under rain applications. - Soil Sci. Soc. Amer. J. 41: 960-966, 1977.

2872 - HADAS, A.: Water uptake and germination of leguminous seeds in soils of changing matric and osmotic water potential. - J. exp. Bot. 28: 977-985, 1977.

2873 - HADLEY, R.F.: Evaluation of land-use and land-treatment practices in semi--arid western United States. - Phil. Trans. roy. Soc. London B 278: 543-554, 1977.

2874 - HAFERKAMP, M.R., JORDAN, G.L., MATSUDA, K.: Physiological development of Lehmann lovegrass seeds during the initial hours of imbibition. Agron. J. 69: 295-299, 1977.

2875 - HAFERKAMP, M.R., JORDAN, G.L., MATSUDA, K.: Pre-sowing seed treatment, seed coats, and metabolic activity of Lehmann lovegrass seeds. - Agron. J. 69: 527-530, 1977.

2876 - HAGON, M.W., CHAN, C.W.: The effects of moisture stress on the germination of some Australian native grass seeds. - Aust. J. exp. Agr. anim. Husb. 17: 86-89, 1977.

2877 - HALL, A.E., THOMSON, W.W., ASBELL, C.W., PLATT-ALOIA, K., LEONARD, R.T.: Stomatal response to humidity and lanthanum. - Physiol. Plant. 41: 89-94, 1977.

2878 - HALLAIRE, M.: Les différents aspects climatiques de la France et leur variation dans le temps: les bilans hydriques en quelques points représentatifs de la France. - Fourrages 1977 (70): 19-35, 1977.

2879 - HANAWALT, R.B., WHITTAKER, R.H.: Altitudinal gradients of nutrient supply to plant roots in mountain soils. - Soil Sci. 123: 85-96, 1977.

2880 - HANCOCK, J.G.: Soluble metabolites in intercellular regions of squash hypocotyl tissues: Implications for exudation. - Plant Soil 47: 103-112, 1977.

2881 - HANDLEY, J.F., JENNINGS, D.H.: The effect of ions on growth and leaf succulence of *Atriplex hortensis* var. cupreata. - Ann. Bot. 41: 1109-1112, 1977.

2882 - HANKS, R.J., SULLIVAN, T.E., HUNSAKER, V.E.: Corn and alfalfa production as influenced by irrigation and salinity. - Soil Sci. Soc. Amer. J.: 41: 606-610, 1977.

2883 - HANKS, R.J., WILLARDSON, L.S., JURINAK, J.J., MELAMED, J.D.: Irrigation management where salinity sources and sinks are present. - In: DREGNE, H.E. (ed.): Managing Saline Water for Irrigation. Pp. 15-27, Texas Technical University, Lubbock 1977.

2884 - HANSCOM III, Z., TING, I.P.: Physiological responses to irrigation in *Opuntia basilaris* Engelm. and Bigel. - Bot. Gaz. 138: 159-167, 1977.

2885 - HANSOM, A.D., NELSEN, C.E., EVERSON, E.H.: Evaluation of free proline accumulation as an index of drought resistance using two contrasting barley cultivars. - Crop Sci. 17: 720-726, 1977.

2886 - HANSON, C.L., RAUZI, F.: Class A pan evaporation as affected by shelter, and a daily prediction equation. - Agr. Meteorol. 18: 27-35, 1977.

2887 - HARDAN, A.: Irrigation with saline water under desert conditions. - In: DREGNE, H.E. (ed.): Managing Saline Water for Irrigation. Pp. 165-169, Texas Technical University, Lubbock 1977.

2888 - HARTGE, K.H., WIEBE, H.-J.: Der Wasserzustand von Pflanze und Boden, sein Einfluss auf die Ertragsbildung und seine Bestimmung. - Gartenbauwissenschaft 42: 71-76, 1977.

2889 - HARVEY, D.M.: Photosynthesis and translocation. - In: SUTCLIFFE, J.F., PATE, J.S. (ed.): The Physiology of Garden Pea. Pp. 315-348. Academic Press, London - New York - San Francisco 1977.

*2890 - HASE, Y., MACHIDA, Y., MAOTANI, T.: [Estimation of the actual evapotranspiration rate by water balance method in deciduous fruit orchard located on a slope.] - Bull. Fruit Tree Res. Sta., Ser. E 1:59-85, 1976. [In Jap, ab: E.]

*2891 - HASTINGS, D.F., GUTKNECHT, J.: Ionic relations and the regulation of turgor pressure in the marine alga, *Valonia macrophysa*. - J. Membrane Biol. 28: 263 - 275, 1976.

2892 - HEATHERLY, L.G., RUSSELL, W.J., HINCKLEY, T.M.: Water relations and growth of soybeans in drying soil. - Crop Sci. 17: 381-386, 1977.

2893 - HÉBANT, C.: The Conducting Tissues of Bryophytes. - J. Cramer, Vaduz 1977.

2894 - HEBBLETHWAITE, P.D.: Irrigation and nitrogen studies in S.23 ryegrass grown for seed. 1. Growth, development, seed yield components and seed yield. - J. agr. Sci. 88: 605-614, 1977.

2895 - HEBBLETHWAITE, P.D., McGOWAN, M.: Irrigation and nitrogen studies in S.23 ryegrass grown for seed. 2. Crop transpiration and soil-water status. - J. agr. Sci. 88: 615-624, 1977.

*2896 - HEGARTY, T.W.: Effects of fertilizer on the seedling emergence of vegetable crops. - J. Sci. Food Agr. 27: 962-968, 1976.

2897 - HEGARTY, T.W.: Seed and seedling susceptibility to phased moisture stress in soil. - J. exp. Bot. 28: 659-668, 1977.

2898 - HEGARTY, T.W.: Seed activation and seed germination under moisture stress. - New Phytol. 78: 349-359, 1977.

2899 - HEILMAN, J.L., KANEMASU, E.T., BAGLEY, J.O., RASMUSSEN, V.P.: Evaluating soil moisture and yield of winter wheat in the Great Plains using Landsat data. - Remote Sensing Environ. 6: 315-326, 1977.

2900 - HELLKVIST, J., PARSBY, J.: The water relations of *Pinus sylvestris*. IV. Diurnal and seasonal patterns of water potential in pine trees from different latitudinal provenances. - Physiol. Plant. 41: 211-216, 1977.

2901 - HELMY, A.K., RON, M.M., FERREIRO, E.A.: Concurrent effects of KCl and pH on root concentration potentials. - Z. Pflanzenphysiol. 81: 269- 277, 1977.

2902 - HERKELRATH, W.N., MILLER, E.E., GARDNER, W.R.: Water uptake by plants: I. Divided root experiments. - Soil Sci. Soc. Amer. J. 41: 1033 -1038. 1977.

2903 - HERKELRATH, W.N., MILLER, E.E., GARDNER, W.R.: Water uptake by plants: II. The root contact model. - Soil Sci. Soc. Amer. J. 41: 1039-1043, 1977.

2904 - HEWETT, E.W.: Sprinkler irrigation of apple trees.-N.Zeal. J. Agr. 134: 23-24, 1977.

2905 - HICKLENTON, P.R., OECHEL, W.C.: The influence of light intensity and temperature on the field carbon dioxide exchange of *Dicranum fuscenscens* n the subarctic. - Arctic alp. Res. 9: 407-419, 1977.

2906 - HIGGINS, D.L., Von BECKMANN, J., JEWELL, E., JOSEPHSON, G.G.S., WILLIS, C.B., SUZUKI, M., THOMPSON, R.G., FENSON, D.S.: Electrical impedance measurements on alfalfa to detect infection by root lesion nematodes. - Can. J. Plant Sci. 57: 853-858, 1977.

*2907 - HIJIMOTO, S., ITANI, T.: [Studies on ecological characteristics of *Solidago altissima* L. 1. Effects of soil moisture, nutrient elements, and shading on the growth.] - Bull.Hiroshima agr. Coll. 5: 117-124, 1975. [In Jap, ab: E.]

2908 - HILLEL, D.: Computer Simulation of Soil-Water Dynamics : A Compendium of Recent Work. - International Development Research Centre, Ottawa 1977.

2909 - HILLEL, D., TALPAZ, H.: Simulation of soil water dynamics in layered soils. - Soil Sci. 123: 54-62, 1977.

*2910 - HILLEL, D., Van BEEK, C.G.E.M., TALPAZ, H.: A microscopic-scale model of soil water uptake and salt movement to plant roots. - Soil Sci. 120: 385-399, 1975.

*2911 - HINDÁK, F.: The phototrophic edaphon of the oak-hornbeam forest at Báb. - In: BISKUPSKÝ, V. (ed.): Research Project Báb, IBP Progress Report II. Pp. 177-183. Veda, Bratislava 1975.

*2912 - HIRANO, J.: Effect of rain in ripening period on the grain quality of wheat. - Jap. agr. Res. Quart. 10: 168-173, 1976.

2913 - HO, L.C., NICHOLS, R.: Translocation of $^{14}$C-sucrose in relation to changes in carbohydrate content in rose corollas cut at different stages of development. - Ann. Bot. 41: 227-242, 1977.

*2914 - HOBBS, E.H., KROGMAN, K.K.: Using evapotranspiration data to favorably influence an irrigated environment. - In: Environmental Impact of Irrigation and Drainage. Pp. 184-191. American Society of Civil Engineers, Ottawa 1976.

2915 - HODGES, C.F.: Influence of irrigation on survival of *Poa pratensis* infected by *Ustilago striiformis* and *Urocystis agropyri*. - Can. J. Bot. 55: 216-218, 1977.

2916 - HOFÄCKER, W.: Untersuchungen zur Stoffproduktion der Rebe unter dem Einfluss wechselnder Bodenwasserversorgung. - Vitis 16: 162-173, 1977.

2917 - HOFER, M.: Transport durch Biologische Membranen. Das Konzept der Trägerkatalyse. - Verlag Chemie, Weinheim - New York 1977.

2918 - HOFFMANN, F.: Untersuchungen zur Ertragsbildung und einigen Bedingungen für die Entstehung hoher Erträge bei Zuckerrüben. - Arch. Acker- Pflanzenbau Bodenk. 21: 157-168, 1977.

2919 - HOFFMANN, P., KRAUSE, C.: Regulative Aspekte der Substanzverteilung auf die einzelnen Organe in Keimpflanzen von *Triticum aestivum* L. im Verlaufe der Entwicklung unter besonderer Berücksichtigung von Energiebilanzen. - In: UNGER, K. (ed.): Biophysikalische Analyse Pflanzlicher Systeme. Pp. 165-174, VEB Gustav Fischer Verlag, Jena 1977.

*2920 - HOFFMANN, P., MICHAELIS, G.: Physiologische Gradienten in Primärblättern von *Triticum aestivum* L. - Wiss. Z. Humboldt-Univ. Berlin, math. - naturwiss. Reihe 25: 787-795, 1976.

2921 - HOFSTRA, J.J., STIENSTRA, A.W.: Growth and photosynthesis of closely related $C_3$ and $C_4$ grasses, as influenced by light intensity and water supply. - Acta bot. Neerl. 26: 63-72, 1977.

2922 - HOLM, L.: Weeds and water in world food production. - Weed Sci. 25: 338-342, 1977.

2923 - HOLSTEIJN, H.M.C., van, BEHBOUDIAN, M.H., BONGERS, H.C.M.L.: Water relations of lettuce. II. Effects of drought on gas exchange properties of two culti- vars. - Scientia Hort. 7: 19-26, 1977.

*2924 - HOLT, D.A., DOUGHERTY, C.T., BULA, R.J., SCHREIBER, M.M., PEART, R.M.: Water relations in SIMED, the purdue model of alfalfa growth. - In: Proceedings of Symposium on Modeling: Climate - Plants - Soils. Pp. 32-45. (University of Guelph, Guelph 1976.

2925 - HOORN, J.W., van, RISSEEUW, J., PRIETO, J.M.J.: Desalinization of a highly sa- line soil, prediction and field observation. - In: DREGNE, H.E. (ed.): Mana- ging Saline Water for Irrigation. Pp. 558-574. Texas Technical University, Lubbock, 1977.

2926 - HOUSLEY, T.L., FISHER, D.B.: Estimation of osmotic gradients in soybean sieve tubes by quantitative autoradiography. Qualified support for Münch hypothesis. - Plant Physiol. 59: 701-706, 1977.

2927 - HUBAC, C., GUERRIER, D., FERRAN, J.: Résultats préliminaires sur le métabo- lisme de la proline en relation avec la résistance à la sécheresse. - C.R. Acad. Sci. Paris, Sér D 284: 1397-1400, 1977.

2928 - HUBER, W., KREUTMEIER, F., SANKHLA, N.: Eco-physiological studies on Indian arid zone plants. VI. Effect of sodium chloride and abscisic acid on amino- -acid and protein metabolism in leaves of *Phaseolus aconitifolius*. - Z. Pflan- zenphysiol. 81: 234-247, 1977.

2929 - HUCK, M.G., KLEPPER, B.: Water relations of cotton. II. Continuous estimates of plant water potential from stem diameter measurements.- Agron. J. 69: 593-597, 1977.

2930 - HUKKERI, S.B., SHUKLA, N.P., RAJPUT, R.K.: Effect of levels of soil moisture and nitrogen on the fodder yield of oat on two types of soils. - Ind. J. agr. Sci. 47: 204-209, 1977.

2931 - HULL, H.M., BLECKMAN, C.A.: An unusual epicuticular wax ultrastructure on lea- ves of *Prosopis tamarugo (Leguminosae)*. - Amer. J. Bot. 64: 1083-1091, 1977.

2932 - HUNGER, W.: Zu den Beziehungen zwischen Wuchsleistung und Ernährungszustand der jüngeren Fichte (*Picea abies* (L.) Karst.) auf Lösslehmstandorten. - Flora 166: 523-536, 1977.

2933 - HUZULÁK, J.: Relative saturation deficit in some forest tree species: develop- ment, daily and seasonal changes. - Biológia (Bratislava) 32: 271-278, 1977.

2934 - HUZULÁK, J.: Diurnal xylem pressure potential patterns in dominant tree spe- cies of oak-hornbeam forest. - Biológia (Bratislava) 32: 469-476, 1977.

2935 - HUZULÁK, J.: Príspevok k štúdiu vodného potenciálu lesných stromov. [A con- tribution to the study of xylem pressure potential in forest trees.] - In: HUZULÁK, J., MASAROVIČOVÁ, E. (ed.): Fotosyntéza a Vodný Režim Drevín. Pp. 63-69. Modra-Piesky 1977. [In Slov, ab: E,R.]

2936 - IARCA, P., VINEŞ, G., LEBEDENCU, I.: Influenţa îngrăşămintelor asupra produc- ţiei de soia în condiţii de irigare pe cernoziomul castaniu din Bărăganul de sud. [The influence of fertilizers upon the soybean yield under irrigation conditions on the chestnut chernozem from the south of Baraganul Plain.] - An. Inst. Cercetări Pentru Cereale Plante Tehnice - Fundulea 42: 163-168, 1977. [In Roum, ab: R,E.]

2937 - IDLE, D.B.: The effects of leaf position and water vapour density deficit on the transpiration rate of detached leaves. - Ann. Bot. 41: 959-968, 1977.

2938 - IDSO, S.B., REGINATO, R.J., JACKSON, R.D.: An equation for potential evapora- tion from soil, water, and crop surfaces adaptable to use by remote sensing. - Geophys. Res. Lett. 4: 187-188, 1977.

2939 - IL'YASHUK, E.M.: Uchastie kaliya v svetoindutsirovannykh dvizheniyakh ust'its u sakharnoi svekly. [Participation of potassium in light-induced movements of stomata in sugar beet.] - Fiziol. Biokhim. kul't. Rast. 9: 285-290, 1977. [In R, ab: E.]

*2940 - IMAM, M.K., ABAZA, M., EL-GAYED, O.: Effect of irrigation and organic manuring
        on growth and yield of Arran Banner potato. - Libyan J. Agr. 4: 79-86, 1975.

2941 -. IMBAMBA, S.K., TIESZEN, L.L.: Influence of light and temperature on photo-
        synthesis and transpiration in some $C_3$ and $C_4$ vegetable plants from Kenya. -
        Physiol. Plant. 39: 311-316, 1977.

2942 - INCOLL, L.D., WHITELAM, G.C.: The effect of kinetin on stomata of the grass
       *Anthephora pubescens* Nees. - Planta 137: 243-245, 1977.

*2943 - INTRIBUS, R.: Water balance factors in the ecosystem of an oak-hornbeam stand
        at the object in Báb. - In: BISKUPSKÝ, V. (ed.): Research Project Báb IBP
        Progress Report II. Pp. 337-352. Veda, Bratislava 1975.

*2944 - ISHIMARU, H.: Crop damage caused by drought. - Jap. agr. Res. Quart. 9: 127-
        130, 1975.

*2945 - ITIER, B., PERRIER, A.: Présentation d'une étude analytique de l'advection.
        I. Advection liée aux  variations horizontales de concentration et de tempé-
        rature. - Ann. agron. 27: 111-140, 1976.

*2946 - ITIER, B., PERRIER, A.: Présentation d'une étude analytique de l'advection.
        II. Application a la mesure et a l'estimation de l'évapotranspiration. -
        Ann. agron. 27: 417-433, 1976.

2947 - IVANCHENKO, V.M., KRUCHININA. S.S., MARSHAKOVA, M.I., URBANOVICH, T.A.:
       K voprosu o mekhanizme fenomena Brilliant. [Mechanism of the Brilliant pheno-
       menon.] - Fiziol. Rast. 24: 416-418, 1977.  [In R.]

*2948 - IVE, J.R., ROSE, C.W., WALL, B.H., TORSSELL, B.W.R.: Estimation and simula-
        tion of sheet run-off. - Aust. J. Soil Res. 14: 129-138, 1976.

2949 - IYENGAR, E.R.R., PATOLIA, J.S., KURIAN, T.: Varietal differences in barley to
       salinity. - Z. Pflanzenphysiol. 84: 355-361, 1977.

2950 - JACKSON, R.D., REGINATO, R.J., IDSO, S.B.: Wheat canopy temperature: a prac-
       tical tool for evaluating water requirements. - Water Resour. Res. 13: 651-
       656, 1977.

2951 - JENÍK, J.: Adaptace k vlhkostnímu stresu u dřevin africké savany. [Adapta-
       tion to moisture stress in woody species of an African savanna.] - In:
       HUZULÁK, J., MASAROVIČOVÁ, Ł. (ed.): Fotosyntéza a Vodný Režim Drevín. Pp.
       77-83. Modra-Piesky 1977. [In Czech, ab: E.]

2952 - JENSEN, C.R.: Effects of salinity in the root medium. IV. Photosynthesis and
       leaf diffusive resistance in relation to $CO_2$-concentration. - Acta Agr. Scand.
       27: 159-164, 1977.

*2953 - JENSEN, H.A.: Investigation of anthesis, length of caryopsis, moisture content,
        seed weight, seed shedding and stripping-ripeness during development and rip-
        ening of a *Festuca pratensis* seed crop. - Acta Agr. Scand. 26: 264-268, 1976.

2954 - JOHNSON, D.A., BROWN, R.W.: Psychrometric analysis of turgor pressure respon-
       se: a possible technique for evaluating plant water stress resistance. - Crop
       Sci. 17: 507-510, 1977.

2955 - JONES, H.G.: Aspects of the water relations of spring wheat (*Triticum aesti-
       vum* L.) in response to induced drought. - J. agr. Sci. 88: 267-282, 1977.

2956 - JONES, H.G.: Transpiration in barley lines with differing stomatal frequencies.
       - J. exp. Bot. 28: 162-168, 1977.

2957 - JONES, H.G., KIRBY, E.J.M.: Effects of manipulation of number of tillers and
       water supply on grain yield in barley. - J. agr. Sci. 88: 391-397, 1977.

2958 - JONES, K.: The effects of moisture on acetylene reduction by mats of blue-
       -green algae in sub-tropical grassland. - Ann. Bot. 41: 801-806, 1977.

2959 - JONES, M.B., DELWICHE, C.C., WILLIAMS, W.A.: Uptake and losses of $^{15}N$ applied
       to annual grass and clover in lysimeters. - Agron. J. 69: 1019-1023, 1977.

2960 - JONES, T.L., JONES, U.S., EZELL, D.O.: Effect of nitrogen and plastic mulch
       on properties of troup loamy sand and on yield of "Walter" tomatoes. - J.
       Amer. Soc. hort. Sci. 102: 273-275. 1977.

2961 - JORDAN, C.F., KLINE, J.R.: Transpiration of trees in a tropical rainforest. -
J. appl. Ecol. 14: 853-860, 1977.

*2962 - JOSHI, G.V.: Photosynthesis under conditions of stress. - Proc. Ind. nat. Sci.
Acad. 42 B: 279-289, 1976.

2963 - JUŘENČÁK, J.: Interakce utuženi a vlhkosti půdy při růstu a vývoji jarního
ječmene. [Interaction of soil compactness and moisture content during the
growth and development of spring barley.] - Rost. Výroba (Praha) 23: 49-58,
1977. [In Czech, ab: R,E,G.]

2964 - JURY, W.A., FLÜHLER, H., STOLZY, L.H.: Problems in predicting and measuring
solute concentrations in irrigated fields. - In: DREGNE, H.E. (ed.): Managing Saline Water for Irrigation. Pp. 246-263. Texas Technical University,
Lubbock, 1977.

2965 - JURY, W.A., FLÜHLER, H., STOLZY, L.H.: Influence of soil properties, leaching
fraction, and plant water uptake on solute concentration distribution. -
Water Resour. Res. 13: 645-650, 1977.

2966 - KAIGAMA, B.K., TEARE, I.D., STONE, L.R., POWERS, W.L.: Root and top growth of
irrigated and nonirrigated grain sorghum. - Crop Sci. 17: 555-559, 1977.

2967 - KALINOWSKA-ZDUN, M.: Sugar beet yielding in the light of the results of the
correlation analyses applied for appretiation of biomass increase in the middle
part of the vegetation period. - In: Produkce Biomasy a Tvorba Výnosu Polních Plodin. Vol. 1. Pp. 136-150. ČVTSZ, Praha 1977.

2968 - KALMAR, D., LAHAV, E.: Water requirements of avocado in Israel. I. Tree and
soil parameters. - Aust. J. agr. Res. 28: 859-868, 1977.

2969 - KANA, T.M., MILLER, J.H.: Effect of colored light on stomatal opening rates
of *Vicia faba* L. - Plant Physiol. 59: 181-183, 1977.

2970 - KANA, T.M., MILLER, J.H.: Effect of photoperiod on stomatal opening in *Vicia
faba*. - Plant Physiol. 60: 803-804, 1977.

*2971 - KANNAN, S., KEPPEL, H.: Intracellular regulation of absorption and transport
of Fe and Mn in wheat seedlings cultured in low and high salt media. - Z.
Pflanzenphysiol. 79: 132-142, 1976.

2972 - KANNANGARA, C.G., GOUGH, S.P.: Synthesis of Δ-aminolevulinic acid and chlorophyll by isolated chloroplasts. - Carlsberg Res. Commun. 42: 441-457, 1977.

2973 - KAPLAN, S.L., KOLLER, H.R.: Leaf area and $CO_2$-exchange rate as determinants
of the rate of vegetative growth in soybean plants. - Crop Sci. 17: 35-38,
1977.

2974 - KARAMI, E.: Effect of irrigation and plant population on yield and yield
components of sunflower. - Ind. J. agr. Sci. 47: 15-17, 1977.

*2975 - KAROLIN, A.Yu., MOLDAU, Kh.A.: Faktorostatnaya kamera s registratsieĭ transpi-
ratsii i $CO_2$-obmena nadzemnykh i podzemnykh chasteĭ rasteniya. [An air-condi-
tioned chamber for recording transpiration and $CO_2$ exchange of aerial and
underground plant organs.] - Fiziol. Rast. 23: 630-634, 1976. [In R, ab: E.]

*2976 - KASZUBIAK, H.: Correlation between determinations of chlorophyll-type com-
pounds, nitrogen available for plants and number of algae in the soil. -
Pol. J. Soil Sci. 9: 47-51, 1976.

2977 - KATERJI, N.B.: Contribution a l'etude de l'evapotranspiration reelle du ble
tendre d'hiver, application a la resistance du couvert en relation avec cer-
tains facteurs du milieu. - These Présente a l'Universite Paris VII pour
Obtenir le Grade de Docteur Ingenieur 1977.

2978 - KATERJI, N.B., GOSSE, G., PERRIER, A., DAUDET, F.A.: Etude suivie de l'évapotranspiration réelle d'un couvert de blé et de maïs au moyen d'un dispositif
automatique B.E.A.R.N. - Meteorologie, Sér. VI e 11: 47-53, 1977.

2979 - KAUFMANN, M.R.: Soil temperature and drying cycle effects on water relations
of *Pinus radiata*. - Can. J. Bot. 55: 2413-2418, 1977.

2980 - KAUFMANN, M.R., ECKARD, A.N.: Water potential and temperature effects on germination of Engelmann spruce and lodgepole pine seeds. - Forest Sci. 23: 27-33, 1977.

2981 - KAUFMANN, M.R., ECKARD, A.N.: A portable instrument for rapidly measuring conductance and transpiration of conifers and other species. - Forest Sci. 23: 227-237, 1977.

2982 - KAUL, R.B.: The role of the multiple epidermis in foliar succulence of *Peperomia (Piperaceae)*. - Bot. Gaz. 138: 213-218, 1977.

2983 - KAUSHAL, M.P., PATHAK, B.S.: Economics and water-use efficiency of sprinkler versus border irrigation on dunes. - Ind. J. agr. Sci. 47: 240-244, 1977.

*2984 - KAUSIK, S.B.: A contribution to foliar anatomy of *Agathis dammara*, with a discussion on the transfusion tissue and stomatal structure. - Phytomorphology 26: 263-273, 1976.

2985 - KAWATA, S., KATANO, M.: [Effect of water management of paddy fields on the direction of crown root growth and the lateral root formation of rice plants.] - Jap. J. Crop Sci. 46: 543-557, 1977. [In Jap, ab: E.]

2986 - KEMP, P.R., WILLIAMS, G.J. III, MAY, D.S.: Temperature relations of gas exchange in altitudinal populations of *Taraxacum officinale*. - Can. J. Bot. 55: 2496-2502, 1977.

2987 - KENNEDY, R.A.: The effect of NaCl-, polyethyleneglycol-, and naturally-induced water stress on photosynthetic products, photosynthetic rates, and $CO_2$ compensation points in $C_4$ plants. - Z. Pflanzenphysiol. 83: 11-24, 1977.

*2988 - KEREČKI, B., ZARIĆ, L., JELENIĆ, D., PENČIĆ, M.: The influence of drought and heat stress on maize plant. - Acta bot. Croat. 34: 188, 1975.

*2989 - KEREČKI, B., ZARIĆ, L., JELENIĆ, D., PENČIĆ, M.: Proučavanje uticaja zemljišne suše i visoke temperature vazduha na neke fiziološke procese u biljci kukuruza. [Study of the influence of soil drought and high air temperature on some physiological processes in maize plant.] - Acta bot. Croat. 35: 113-118, 1976. [In Croat, ab: E.]

2990 - KERSHAW, K.A., DZIKOWSKI, P.A.: Physiological-environmental interactions in lichens. - VI. Nitrogenase activity in *Peltigera polydactyla* after a period of desiccation. - New Phytol. 79: 417-421, 1977.

2991 - KHAVARI-NEJAD, R.A.: Effects of α-hydroxy-2-pyridinemethanesulfonic acid on photosynthetic carbon dioxide uptake and stomatal movements in excised tomato leaves. - Plant Physiol. 60: 44-46, 1977.

2992 - KHOKHLOVA, L.P., BONDAR', I.G., ELISEEVA, N.S., PANKRATOVA, S.I., SULEĬMANOV, I.G.: Izuchenie sostoyaniya vody v uzlakh kushneniya i izolirovannykh iz nikh mitokhondriyakh ozimykh pshenits metodom yadernogo magnitnogo rezonansa. [The state of water in tillering nodes of winter wheat and mitochondria isolated from them studied by NMR.] - Fiziol. Rast. 24: 620-626, 1977. [In R, ab: E.]

*2993 - KHRISTOV, Kh.D., PETROVA, L.I.: Vliyanie na giberelinovata kiselina v"rkhu s"stoyanieto na vodata v prorast"tsi ot *Salvia*, otglezhdani na t"mno i na svetlo. [Effect of gibberellic acid on the water status of *Salvia* seedlings grown in the dark and in the light.] - Fiziol. Rast. (Sofia) 11: 79-87, 1976. [In Bulg, ab: E.]

2994 - KING, R.W., EVANS, L.T.: Inhibition of flowering in *Lolium temulentum* L. by water stress: a role for abscisic acid. - Aust. J. Plant Physiol. 4: 225-233, 1977.

*2995 - KIRIK, N.N., STEBLYUK, N.I., ÉLLANSKAYA, I.A.: Morfologicheskie i biologicheskie osobennosti vozbuditeleĭ fuzarioznoĭ kornevoĭ gnili i uvyadaniya gorokha. [Morphological and biological characteristics of the pathogens of fusarious root rot and wilt in pea.] - Sel'skokhoz. Biol. 11: 689-694, 1976. [In R, ab: E.]

2996 - KIRST, G.O.: The cell volume of the unicellular alga, *Platymonas subcordiformis* Hazen: Effect of the salinity of the culture media and of osmotic stresses. - Z. Pflanzenphysiol. 81: 386-394, 1977.

2997 - KIRST, G.O.: Coordination of ionic relations and mannitol concentrations in the euryhaline unicellular alga, *Platymonas subcordiformis* (Hazen) after osmotic shocks. - Planta 135: 69-75, 1977.

2998 - KLISIEWICZ, J.M.: Effect of flooding and temperature on incidence and severity of safflower seedling rust and viability of *Puccinia carthami* teliospores. - Phytopathology 67: 787-790, 1977.

2999 - KNAUF, T.A., BILAN, M.V.: Cotyledon and primary needle variation in loblolly pine from mesic and xeric seed sources. - Forest Sci. 23: 33-36, 1977.

3000 - KNEDLHANS, S.: A method for recording evaporation. - Agr. Meteorol. 18: 487-489, 1977.

3001 - KNOF, G.: Eine transportable Messanordnung zur Erfassung der $CO_2$-Aufnahme, der Beleuchtungsstärke, der Temperatur und des relativen Wassergehaltes an Pflanzenblättern in Feldbeständen. - Arch. Acker- Pflanzenbau Bodenk. 21: 35-44, 1977.

*3002 - KONONOVICH, A.I.: Ustoĭchivost' soi k zasukhe i izbytochnomu uvlazhneniyu pochv v usloviyakh Amurskoĭ oblasti. [Resistance of soybean to drought and to waterlogging in region of Amur river.] - In: Ustoĭchivost' Rasteniĭ k Pereuvlazhneniyu Pochvy v Usloviyakh Dal'nego Vostoka. Pp. 83-94. DVGU, Vladivostok 1976. [In R.]

*3003 - KONONOVICH, A.I., GONTA, V.S.: Reaktsiya razlichnykh sortov soi na pereuvlazhnenie pochvy i otzyvchivost' ikh na udobreniya. [Reaction of different soybean varieties on waterlogging and mineral nutrition.] - In: Ustoĭchivost' Rasteniĭ k Pereuvlazhneniyu Pochvy v Usloviyakh Dal'nego Vostoka. Pp. 95-101. DVGU, Vladivostok 1976. [In R.]

3004 - KOO, R.C.J., REESE, R.L.: Fertility and irrigation effects on "Temple" orange. II. Fruit quality. - J. Amer. Soc. hort. Sci. 102: 152-155, 1977.

3005 - KÖRNER, C.: Evapotranspiration und Transpiration verschiedener Pflanzenbestände im alpinen Grasheidegürtel der Hohen Tauern. - Veröf. Österreich. Mass-Hochgebirgsprogrammes Hohe Tauern 1: 47-68, 1977.

3006 - KÖRNER, C.: Blattdiffusionswiderstände verschiedener Pflanzen im alpinen Grasheidegürtel der Hohen Tauern. - Veröf. Österreich. Mass-Hochgebirgsprogrammes Hohe Tauern 1: 69-81, 1977.

*3007 - KORSHUNOV, A.V., POPOV, B.A.: Vysokie urozhai kartofelya: optimal'nye normy udobreniĭ i oroshenie. [High yield of potato: optimal fertilization and irrigation.] - Vestn. sel'skokhoz. Nauki 1976 (2): 92-96, 1976. [In R.]

*3008 - KOSMAKOVA, V.E., ZVEREVA, E.G.: Aktivnost' fotosinteticheskogo apparata rasteniĭ soi v usloviyakh pereuvlazhneniya pochvy. [Activity of photosynthetic apparatus of soybean plants under conditions of waterlogging.] - In: Ustoĭchivost' Rasteniĭ k Pereuvlazhneniyu Pochvy v Usloviyakh Dal'nego Vostoka. Pp. 3-25. DVGU, Vladivostok 1976. [In R.]

3009 - KOTSYUBYNS'KA, N.P.: Vytraty vody na transpiratsiyu golovnymy licoutvoryuyuchimi porodami v dibrovnykh fitotsenozakh stepovoĭ Ukrainy. [Water expenditure for transpiration by main forest formation species in the oak forest phytocenoses of the steppe in Ukraine.] - Ukr. bot. Zh. 34: 119-122, 1977. [In Ukr, ab: E.]

3010 - KOZINKA, V.: Koreň rastlín ako orgán pozdĺžného transportu vody. [The root as the organ of longitudinal water transport in plants.] - In: Dni Rastlinnej Fyziológie I. Pp. 164-167, 307. Slovenská Botanická Spoločnost SAV, Bratislava 1977. [In Slov, ab: R,E.]

*3011 - KOZLOWSKI, T.T.: Drought and transplantability of trees. - USDA Forest Serv. Gen. Tech. Rep. NE-22: 77-90, 1976.

3012 - KRAUSE, D., KUMMEROW, J.: Xeromorphic structure and soil moisture in chaparral. - Oecol. Plant. 12: 133-148, 1977.

3013 - KREEB, K.: Methoden der Pflanzenökologie. - VEB Gustav Fischer Verlag, Jena 1977.

*3014 - KROGMAN, K.K., HOBBS, E.H.: Scheduling irrigation to meet crop demands. - Agr. Canada Publ. 1590: 1-18, 1976.

3015 - KROGMAN, K.K., HOBBS, E.H.: Irrigation management of alfalfa for seed. - Can. J. Plant Sci. 57: 891-896, 1977.

*3016 - KRÓLIKOWSKA, J.: Physiological effect of triazine herbicides on *Typha latifolia* L. - Pol. Arch. Hydrobiol. 23: 249-259, 1976.

*3017 - KRULIKOVSKA, E.: Vliyanie linurona na transpiratsiyu rogoza *Typha latifolia* L. [The effect of linuron on transpiration of *Typha latifolia*.] - Éksperiment. vodn. Toksikol. 6: 54-64, 1976. [In R, ab: E.]

3018 - KRUPA, J.: The interdependence between transpiration intensity and the anatomical structure of moss leaves. - Acta Soc. bot. Pol. 46: 57-68, 1977.

3019 - KRŮŽELA, J.: Dynamika tvorby kořenů řepky ozimé při rozdílné vlhkosti půdy. [Dynamics of root formation in winter rape under different soil moisture content.] - Rost. Výroba (Praha) 23: 439-448, 1977. [In Czech, ab: E,R,G.]

3020 - KU, S.-B., EDWARDS, G.E., TANNER, C.B.: Effect of light, carbon dioxide, and temperature on photosynthesis, oxygen inhibition of photosynthesis, and transpiration in *Solanum tuberosum*. - Plant Physiol. 59: 868-872, 1977.

3021 - KU, S.-B., HUNT, L.A.: Effects of temperature on the photosynthesis-irradiance response curves of newly matured leaves of alfalfa. - Can. J. Bot. 55: 872-879, 1977.

3022 - KUČERA, J., ČERMÁK, J., PENKA, M.: Improved thermal method of continual recording the transpiration flow rate dynamics. - Biol. Plant. 19: 413-420, 1977.

3023 - KUMAKHOVA, T.A., JATSENKO, L.M.: Soderzhanie svobodnykh aminokislot v list'yakh ozimoǐ pshenitsy pri razlichnykh usloviyakh vodoobespechennosti i pitaniya. [Content of free amino acids in winter wheat leaves under different conditions of water supply and nutrition.] - Fiziol. Biokhim. kul't. Rast. 9: 588-591, 1977. [In R, ab: E.]

*3024 - KUMAR, N.C.: Eco-physiological studies of *Dendrophthoe falcata* infection. - Bull. bot. Soc. Bengal 29: 33-38, 1975.

*3025 - KUMASHIRO, K., SATO, Y., TATEISHI, S.: [Studies on the leaf burn in *Pyrus* spp. III. Relationship between the resistance to the leaf burn and the drought resistance of detached leaves.] - J. Jap. Soc. hort. Sci. 43: 377-382, 1975. [In Jap, ab: E.]

*3026 - KURAISHI, S.: Ineffectiveness of cytokinin-induced chlorophyll retention in hypostomatous leaf discs. - Plant Cell Physiol. 17: 875-885, 1976.

3027 - KURAISHI, S., ISHIKAWA, F.: Relationship between transpiration and amino acid accumulation in *Brassica* leaf discs treated with cytokinins and fusicoccin. - Plant Cell Physiol. 18: 1273-1279, 1977.

3028 - KUSHNIRENKO, M.D., PECHERSKAYA, S.N., KRYUKOVA, E.V., KANASH, E.V.: Vliyanie vlazhnosti sredy na pigment-belkovyǐ kompleks list'ev i khloroplastov grushi. [Effect of medium moisture on the pigment-protein complex in pear-tree leaves and chloroplasts.] - Fiziol. Biokhim. kul't. Rast. 9: 625-630, 1977. [In R, ab: E.]

3029 - KYRIAKOPOULOS, E., RICHTER, H.: A comparison of methods for the determination of water status in *Quercus ilex* L. - Z. Pflanzenphysiol. 82: 14-27, 1977.

*3030 - LAD, S.L., KALBHOR, P.N.: To study the effects of irrigation scheduled according to water use factors on the yield and yield contributing characters of wheat (*Triticum aestivum* L.) var. Nl. 747-19 under varying fertility levels of nitrogen and phosphate. - J. Maharashtra agr. Univ. 1: 234-237, 1976.

*3031 - LAD, S.L., KALBHOR, P.N.: Effects of irrigation scheduled according to water use factors on soil moisture use pattern and water use efficiency of wheat (*Triticum aestivum* L.), var. Nl. 747-19. - J. Maharashtra agr. Univ. 1: 237-242, 1976.

3032 - LAD, S.L., KALBHOR, P.N.: Effects of irrigation scheduled according to water use factors on soil moisture content and consumptive use of water by wheat

(*Triticum aestivum* L.) var. NI. 747-19. - J. Maharashtra agr. Univ. 2: 127-134, 1977.

3033 - LAD, S.L., KALBHOR, P.N.: Nitrogen and phosphate uptake by wheat (*Triticum aestivum* L.) var. NI.747-19 under irrigation and varying levels of nitrogen and phosphate. - J. Maharashtra agr. Univ. 2: 168-170, 1977.

3034 - LADIGES, P.Y., KELSO, A.: The comparative effects of waterlogging on two populations of *Eucalyptus viminalis* Labill. and one population of *E. ovata* Labill. - Aust. J. Bot. 25: 159-169, 1977.

3035 - LAHAV, E., KALMAR, D.: Water requirements of avocado in Israel. II. Influence on yield, fruit growth and oil content. - Aust. J. agr. Res. 28: 869-877, 1977.

*3036 - LAL, R.: Soil management to increase water and fertilizer efficiency. - In: Efficiency of Water and Fertilizer Use in Semi-Arid Regions. A Technical Document. Pp. 101-122. International Atomic Energy Agency, Vienna 1976.

*3037 - LAL, R.B., BAINS, S.S., RAY, S.: Effect of methods of irrigation on the yield and water use of different crops under different cropping systems. - Ind. J. agr. Sci. 45: 410-421, 1975.

*3038 - LAMBERT, J.P., RUMBALL, P.J., CHRISTIE, A.J.R.: Comparison of ryegrass and kikuyu grass pastures under mowing. - N.Zeal. J. exp. Agr. 5: 71-77, 1976.

3039 - LAMONT, B., PERRY, M.: The effects of light, osmotic potential and atmospheric gases on germination of the mistletoe *Amyema preisii*. - Ann. Bot. 41: 203-209, 1977.

3040 - LANCASTER, J.E., MANN, J.D., PORTER, N.G.: Ineffectiveness of abscisic acid in stomatal closure of yellow lupin, *Lupinus luteus*, var. Weiko III. - J. exp. Bot. 28: 184-191, 1977.

3041 - LANDSBERG, J.J., CUTTING, C.V. (ed.): Environmental Effects on Crop Physiology. (Proccedings of a Symposium held at Long Ashton Research Station University Bristol 13 -16 April 1975.) Academic Press, London - New York - San Francisco 1977.

3042 - LANGE, O.L., GEIGER, I.L., SCHULZE, E.-D.: Ecophysiological investigations on lichens of the Negev desert. V. A model to simulate net photosynthesis and respiration of *Ramalina maciformis*. - Oecologia 28: 247-259, 1977.

*3043 - LAPINSKENE, N.A., GUTAUSKAS, L.V.: Podzemnaya chast' travosmeseĭ oroshaemykh pastbishch kholmistogo rel'efa Litovskoĭ SSR. [Underground part of grass medley in irrigated pastures of the hilly relief of the Lithuanian SSR.] - Tr. Akad. Nauk Litovskoĭ SSR, Ser. B 2(74): 3-14, 1976. [In R, ab: Lithu,E.]

*3044 - LAPTEV, Yu.P., MAKAROV, P.P., GLAZOVA, M.V., SHUGAEVA, E.V., MIKHAĬLOVA, S.P., ARKHANGEL'SKAYA, M.A., VLADIMIROVA, I.A.: Ust'ichnyĭ apparat i pyl'tsa kak pokazateli ploidnosti rasteniĭ. [Stomatal apparatus and pollen as indicators of plant ploidy.] - Genetika 12(1): 47-55, 1976. [In R, ab: E.]

3045 - LARSON, D.W.: A method for the in situ measurement of lichen moisture content. - J. Ecol. 65: 135-145, 1977.

3046 - LARSON, P.R.: Phyllotactic transitions in the vascular system of *Populus deltoides* Bartr. as determined by $^{14}C$ labeling. - Planta 134: 241-249, 1977.

3047 - LASSOIE, J.P., FETCHER, N., SALO, D.J.: Stomatal infiltration pressures versus diffusion porometer measurements of needle resistance in Douglas-fir and lodgepole pine foliage. - Can. J. Forest Res. 7: 192-196, 1977.

3048 - LASSOIE, J.P., SCOTT, D.R.M., FRITSCHEN, L.J.: Transpiration studies in Douglas-fir using the heat pulse technique. - Forest Sci. 23: 377-390, 1977.

3049 - LAWLOR, D.W., FOCK, H.: Photosynthetic assimilation of $^{14}CO_2$ by water-stressed sunflower leaves at two $O_2$ concentrations and the specific activity of products. - J. exp. Bot. 28: 320-328, 1977.

3050 - LAWLOR, D.W., FOCK, H.: Water stress induced changes in the amounts of some photosynthetic assimilation products and respiratory metabolites of sunflower leaves. - J. exp. Bot. 28: 329-337, 1977.

3051 - LAWN, R.J., BYTH, D.E., MUNGOMERY, V.E.: Response of soybeans to planting date in south-eastern Queensland. III. Agronomic and physiological response of cultivars to planting arrangements. - Aust. J. agr. Res. 28: 63-79, 1977.

*3052 - LAZAROV, R., MEKHANDZHIEVA, A., UG"RCHINSKI, S.: Napoyavane na tsarevitsata s namaleni napoitelni normi. [Irrigating maize at reduced seasonal depths of irrigation water.] - Rasteniev"dni Nauki 13: 40-50, 1976. [In Bulg, ab: R, E.]

3053 - LEA, H.Z., DUNN, G.M., KOCH, D.W.: Stomatal diffusion resistance in three ploidy levels of smooth bromegrass. - Crop Sci. 17: 91-93, 1977.

3054 - LEA, H.Z., DUNN, G.M., KOCH, D.W.: Stomatal indexes in three ploidy levels of *Bromus inermis* Leyss. - Crop Sci. 17: 669-670, 1977.

*3055 - LEBEDEV, G.V., EGOROV, V.G., BRYUKVIN, V.G., SABININA, E.D.: Impul'snoe Dozhdevanie Rasteniĭ. Teoriya i Praktika. [Sprinkling Irrigation of Plants. Theory and Practice.] - Nauka, Moskva 1976. [In R.]

3056 - LECHEVALLIER, D.: Effets du polyéthylène glycol sur les lipides et les lipochromes des colonies de Spirodèle. - Physiol. vég. 15: 387-402, 1977.

3057 - LEDENT, J.F.: Effect of partial defoliation and vein cutting on grain growth and yield in winter wheat (*Triticum aestivum* L.). - Bull. Soc. roy. Bot. Belg. 110: 239-250, 1977.

*3058 - LEMOINE-SÉBASTIAN, C.: Structures épidermiques chez *Cryptomeria japonica* (L.) Don. et nature de l'ecaille séminale. - In: Actes du 97$^e$ Congrés National des Sociétés Savantes, Nantes 1972. Section des Sciences, Tome IV. Pp. 407-419. Bibliothéque Nationale, Paris 1976.

3059 - LEVI, I., BERNER, T., COHEN, Y.: Primary production in loess soil crusts of the Negev. - Isr. J. Bot. 26: 44, 1977.

3060 - LEVITT, J.: Effect of environmental stress on transport of ions across membranes. - In: MARRÈ, E., CIFERRI, O. (ed.): Regulation of Cell Membrane Activities in Plants. Pp. 103-119. Elsevier/North-Holland Biomedical Press, Amsterdam - Oxford - New York 1977.

*3061 - LEVY, Y., KAUFMANN, M.R.: Cycling of leaf conductance in citrus exposed to natural and controlled environments. - Can. J. Bot. 54: 2215-2218, 1976.

3062 - LEWIS, D.A., NOBEL, P.S.: Thermal energy exchange model and water loss of a barrel cactus, *Ferocactus acanthodes*. - Plant Physiol. 60: 609-616, 1977.

*3063 - LINDNER, K., ROGASIK, H., LINDNER, H.: Zur Wirkung von Stauschichten auf die Wasserspeicherung leichter Sandböden und den Ertrag. - Arch. Acker- Pflanzenbau Bodenk. 19: 695-709, 1975.

3064 - LOACH, K.: Leaf water potential and the rooting of cuttings under mist and polythene. - Physiol. Plant. 40: 191-197, 1977.

3065 - LOCÁSCIO, S.J., MYERS, J.M., MARTIN, F.G.: Frequency and rate of fertilization with trickle irrigation for strawberries. J. Amer. Soc. hort. Sci. 102: 456-458, 1977.

*3066 - LOCH, D.S., HOPKINSON, J.M., ENGLISH, B.H.: Seed production of *Stylosanthes guyanensis*. 3. Effects of pre-harvest desiccation. - Aust. J. exp. Agr. anim. Husb. 16: 231-233, 1976.

*3067 - LOF, H.: Water use efficiency and competition between arid zone annuals, especially the grasses *Phalaris minor* and *Hordeum murinum*. - Agr. Res. Rep. (Wageningen) 853: 1-107, 1976.

3068 - LONGSTRETH, D.J., STRAIN, B.R.: Effects of salinity and illumination on photosynthesis and water balance of *Spartina alterniflora* Loisel. - Oecologia 31: 191-199, 1977.

*3069 - LORESTO, G.C., CHANG, T.T., TAGUMPAY, O.: Field evaluation and breeding for drought resistance. - Philippine J. Crop Sci. 1: 36-39, 1976.

3070 - LÖSCH, R.: Responses of stomata to environmental factors - experiments with isolated epidermal strips of *Polypodium vulgare*. I. Temperature and humidity. - Oecologia 29: 85-97, 1977.

3071 - ŁOTOCKI, A.: Effect of root aeration and form of nitrogen on photosynthetic
       productivity of Scots pine (*Pinus silvestris* L.) - Acta Soc. Bot. Pol. 46:
       303-316, 1977.

3072 - LOUWERSE, W., ZWEERDE, W.V.D.: Photosynthesis, transpiration and leaf morphol-
       ogy of *Phaseolus vulgaris* and *Zea mays* grown at different irradiances in arti-
       ficial and sunlight. - Photosynthetica 11: 11-21, 1977.

3073 - LOVEYS, B.R.: The intracellular location of abscisic acid in stressed and
       non-stressed leaf tissue. - Physiol. Plant. 40: 6-10, 1977.

3074 - LUCA, P. de, ALFANI, A., VIRZO de SANTO, A.: CAM, transpiration, and adaptive
       mechanisms to xeric environments in the succulent *Cucurbitaceae*. - Bot. Gaz.
       138: 474-478, 1977.

3075 - LUDLOW, M.M., NG, T.T.: Leaf elongation rate in *Panicum maximum* var. *tricho-
       glume* following removal of water stress. - Aust. J. Plant Physiol. 4: 263-
       272, 1977.

3076 - LURIE, S.: Stomatal development in etiolated *Vicia faba:* relationship between
       structure and function. - Aust. J. Plant Physiol. 4: 61-68, 1977.

3077 - LURIE, S.: Stomatal opening and photosynthesis in greening leaves of *Vicia
       faba* L. - Aust. J. Plant Physiol. 4: 69-74, 1977.

3078 - LÜTTGE, U., BALL, E.: Concentration and pH dependence of malate efflux and
       influx in leaf slices of CAM plants. - Z. Pflanzenphysiol. 83: 43-54, 1977.

3079 - LÜTTGE, U., BALL, E.: Water relation parameters of the CAM plant *Kalanchoë
       daigremontiana* in relation to diurnal malate oscillations. - Oecologia 31: 85-
       94, 1977.

3080 - LÜTTGE, U., BALL, E., GREENWAY, H.: Effects of water and turgor potential on
       malate efflux from leaf slices of *Kalanchoë daigremontiana*. - Plant Physiol.
       60: 521-523, 1977.

3081 - LUXMOORE, R.J., ROOYEN, D.J. van, HOLE, F.D., MANKIN, J.B., GOLDSTEIN, R.A.:
       Field water balance and simulated water relations of prairie and oak-hickory
       vegetation on deciduous forest soils. - Soil Sci. 123: 77-84, 1977.

3082 - LYSHEDE, O.B.: Structure of the epidermal and subepidermal cells of some de-
       sert plants of Israel. *Anabasis articulata* and *Calligonum comosum*. - Isr. J.
       Bot. 26: 1-10, 1977.

3083 - LYSHEDE, O.B.: Anatomical features of some stem assimilating desert plants of
       Israel. - Bot. Tidsskr. 71: 225-230, 1977.

3084 - MAAS, E.V., HOFFMAN, G.J.: Crop salt tolerance - current assessment. - J.
       Irrig. Drain. Div. ASCE 103: 115-134, 1977.

3085 - MAAS, E.V., HOFFMAN, G.J.: Crop Salt tolerance: Evaluation of existing data. -
       In: DREGNE, H.E. (ed.): Managing Saline Water for Irrigation. Pp. 187-198.
       Texas Technical University, Lubbock 1977.

*3086 - MACHIDA, Y., HASE, Y., MAOTANI, T., YAMATSU, K., YAMASAKI, T.: [Evapotranspi-
       ration from a satsuma mandarin orchard determined by energy balance method.]
       - Bull. Fruit Tree Res. Sta., Ser. E (Akitsu) 1: 37-50, 1976. [In Jap, ab: E.]

3087 - MADER, P., CHLAD, F., CHLADOVÁ, J., NAUŠ, J., KOHLOVÁ, V., SOFROVÁ, D., MAKO-
       VEC, P.: Development of photosynthetic apparatus of etiolated seedlings of
       *Hordeum vulgare* grown under different nutrition conditions and water state. -
       In: Pigment-Protein Complexes in Photosynthesis. Biological Research Center,
       Szeged 1977.

*3088 - MAGRISO, Yu., SLAVCHEVA, T.: Vliyanie na pochveneta vlaga i toreneto v"rkhu
       intenziteta na fotosintezata na lozata. [Effect of soil moisture and nutrient
       level on photosynthesis of *Vitis vinifera*.] - Gradinarska lozarska Nauka 12:
       64-73, 1975. [In Bulg, ab: R, F.]

*3089 - MAGRISO, Yu., SLAVCHEVA, T.: Vliyanie na pochvenata vlaga v"rkhu intenziteta
       na fotosintezata i transpiratsiyata pri lozata. [Influence of soil humidity on
       photosynthesis and transpiration of *Vitis vinifera*.] - Gradinarska lozarska
       Nauka 13: 72-80, 1976. [In Bulg, ab: R, F.]

*3090 - MAHALL, B.E.: Possible use of hollow fiber hemodialysers in soil-plant water relations research.- Carnegie Inst. Year Book 75: 438-440, 1976.

3091 - MAHALL, B.E., PARK, R.B.: The ecotone between *Spartina foliosa* Trin. and *Salicornia virginica* L. in salt marshes of northern San Francisco bay. II. Soil water and salinity. - J. Ecol. 64: 793-809, 1977.

3092 - MAHON, J.D., LOWE, S.B., HUNT, L.A.: Variation in the rate of photosynthetic $CO_2$ uptake in cassava cultivars and related species of *Manihot*. - Photosynthetica 11: 131-138, 1977.

3093 - MAHON, J.D., LOWE, S.B., HUNT, L.A., THIAGARAJAH, M.: Environmental effects on photosynthesis and transpiration in attached leaves of cassava (*Manihot esculenta* Crantz.) - Photosynthetica 11: 121-130, 1977.

*3094 - MAĬCHEKINA, R.M.: Anatomiya fotosinteziruyushchikh organov ozimykh pshenits v svyazi s sortovymi osobennostyami rasteniĭ. [Anatomy of photosynthesizing organs of winter wheat in relation to cultivar characteristics of plants.] - In: Fotosintez i Produktivnost' Ozimoĭ Pshenitsy na Yugo-Vostoke Kazakhstana. Pp. 114-121, 133-134. Nauka Kazakh.SSR, Alma-Ata 1976. [In R.]

3095 - MALI, C.V., VARADE, S.B., MUSANDE, V.G.: Note on water diffusivity in cotton seed during germination. - Ind. J. agr. Sci. 47: 582-584, 1977.

3096 - MALI, P.C., MEHTA, S.L.: Effect of drought on enzymes and free proline in rice varieties. - Phytochemistry 16: 1355-1358, 1977.

*3097 - MALLICK, S., RAO, T.V., NAGARAJARAO, Y.: Effect of sub-surface compaction and bentonite application on the irrigation reguirement and growth of rice. - Ind. J. Agron. 21: 317-318, 1976.

*3098 - MALOFEEV, V.M.: Samoregulyatsiya fotosinteticheskoĭ funktsii rasteniĭ pri rezkoĭ smene pogody v polevykh usloviyakh. [Self-regulation of photosynthetic function in plants after a rapid change in weather in field conditions.] - Dokl. TSKhA (Moskva) 209: 5-9, 1975. [In R.]

3099 - MALOFEEV, V.M.: Odnovremennaya i nepreryvnaya registratsiya $O_2$ i $CO_2$ pri issledovanii nestatsionarnykh sostoyaniĭ fotosinteza rasteniĭ. [Simultaneous and continuous registration of oxygen and carbon dioxide during studies of non-steady state photosynthesis of plants.] - Fiziol. Rast. 24: 203-206, 1977. [In R, ab: E.]

3100 - MANAM, R., TEARE, I.D., POWERS, W.L., SKIDMORE, E.L.: Nitrate reductase activity of soybeans in relation to other indicators of water stress. - Fyton 35: 189-194, 1977.

*3101 - MANCHANDA, H.R., BHANDARI, D.K.: Effect of pre-soaking of seeds in salt solutions on the yield of wheat and barley irrigated with highly saline waters. - J. Ind. Soc. Soil Sci. 24: 432-435, 1976.

3102 - MANNING, C.E., MILLER, D.G., TEARE, I.D.: Effect of moisture stress on leaf anatomy and water use efficiency of peas. - J. Amer. Soc. hort. Sci. 102: 756-760, 1977.

3103 - MANOHAR, M.S.: A versatile Peltier psychrometer and evaluation of leaf water potential and its components during water stress cycles in some species. - Z. Pflanzenphysiol. 84: 147-158, 1977.

3104 - MANOHAR, M.S.: Gradients of water potential and its components in leaves and shoots of eucalyptus. - Z. Pflanzenphysiol. 84: 227-235, 1977.

3105 - MANOLAKIS, E., LÜDDERS, P.: Die Wirkung gleichmässiger und jahreszeitlich abwechselnder Ammonium- und Nitraternährung auf Apfelbäume. II. Einfluss auf den Wasserverbrauch und die Nährstoffaufnahme. - Gartenbauwissenschaft 42: 79-87, 1977.

*3106 - MANSFIELD, T.A. (ed.): Effects of Air Pollutants on Plants. (Society for Experimental Biology - Seminar Series 1.). Cambridge University Press, Cambridge - London - New York - Melbourne 1976.

3107 - MAOTANI, T., MACHIDA, Y.: [Studies on leaf diffusion resistance of fruit trees. I. Methods of measuring leaf diffusion resistance of satsuma mandarin trees and factors influencing it.] - J. Jap. Soc. hort. Sci. 46: 1-8, 1977. [In Jap. ab: E.]

3108 - MAOTANI, T., MACHIDA, Y.: [Studies on leaf water stress in fruit trees. VII. Effects of summer water potential of satsuma mandarin trees on fruit characteristics at harvest time.] - J. Jap. Soc. hort. Sci. 46: 145-152, 1977. [In Jap, ab: E.]

3109 - MAOTANI, T., MACHIDA, Y.: [Changes in transpiration rate, leaf diffusion resistance and leaf water potential for satsuma mandarin (*Citrus unshiu* Marc.) trees during prolonged water stress and subsequent recovery.] - J. agr. Meteorol. 32: 203-208, 1977. [In Jap, ab: E.]

3110 - MAOTANI, T., MACHIDA, Y., YAMATSU, K.: [Studies on leaf water stress in fruit trees. VI. Effect of leaf water potential on growth of satsuma mandarin (*Citrus unshiu* Marc.) trees.] - J. Jap. Soc. hort. Sci. 45: 329-334, 1977. [In Jap, ab: E.]

3111 - MARCELLOS, H.: Wheat frost injury - freezing stress and photosynthesis. - Aust. J. agr. Res. 28: 557-564, 1977.

*3112 - MARETZKI, A., THOM, M., MOORE, P.H.: Growth patterns and carbohydrate distribution in sugarcane plants treated with an amine salt of glyphosate. - Hawaiian Planters' Record 59: 21-32, 1976.

3113 - MARGARIS, N.S.: Water relations in plants dominating phryganic ecosystems. - Biol Plant. 19: 442-447, 1977.

3114 - MARRÉ, E.: Physiologic implications of the hormonal control of ion transport in plants. - In: PILET, P.E. (ed.) Plant Growth Regulation. Pp. 54-66. Springer-Verlag, Berlin - Heidelberg - New York 1977.

3115 - MARSHALL, P.E., KOZLOWSKI, T.T.: Changes in structure and function of epigeous cotyledons of woody angiosperms during early seedling growth. - Can. J. Bot. 55: 208-215, 1977.

3116 - MARTIN, J.K.: Effect of soil moisture on the release of organic carbon from wheat roots. - Soil Biol. Biochem. 9: 303-304, 1977.

3117 - MARX, A., SACHS, T.: The determination of stomata pattern and frequency in *Anagallis*. - Bot. Gaz. 138: 385-392, 1977.

3118 - MATAR, A.E.: Yields and response of cereal crops to phosphorus fertilization under changing rainfall conditions. - Agron. J. 69: 879-881, 1977.

3119 - MATHERS, A.C., STEWART, B.A., THOMAS, J.D.: Manure effects on water intake and runoff quality from irrigated grain sorghum plots. - Soil Sci. Soc. Amer. J. 41: 782-785, 1977.

3120 - MATHRE, D.E., JOHNSTON, R.H.: Physical and chemical factors affecting sporulation of *Hymenula cerealis*. - Trans. Brit. mycol. Soc. 69: 213-216, 1977.

*3121 - MATTHEWS, S., COLLINS, M.T.: Laboratory measures of field emergence potential in barley. - Seed Sci. Technol. 3: 863-870, 1975.

*3122 - MAVI, H.S., NEWMAN, J.E.: Crop production potentials in India - a water-availability based analysis. - Agr. Meteorol. 17: 387-395, 1976.

3123 - MAYR, H.H., PRESOLY, E., RITTMEYER, G.: Auswirkungen von Behandlungen mit Fettsäurederivaten auf Wasserhaushalt und Ertrag verschiedener Kulturpflanzen. - Z. Pflanzenernähr. Bodenk. 140: 463-472, 1977.

*3124 - McBRIDE, J.R., STONE, E.C.: Plant succession on the sand dunes of the Monterey Peninsula, California. - Amer. Midl. Nat. 96: 118-132, 1976.

3125 - McCOLL, J.G.: Retention of soil water following forest cutting. - Soil Sci. Soc. Amer. J. 41: 984-988, 1977.

3126 - McCREE, K.J., VAN BAVEL, C.H.M.: Calibration of leaf resistance porometers. - Agron. J. 69: 724-726, 1977.

3127 - McCREE, K.J., VAN BAVEL, C.H.M.: Respiration and crop production: A case study with two crops under water stress. - In: LANDSBERG, J.J., CUTTING, C.V. (ed.): Environmental Effects on Crop Physiology. Pp. 199-216. Academic Press, London - New York - San Francisco 1977.

3128 - McINTYRE, G.I.: The role of nutrition in apical dominance. - In: JENNINGS, D.H. (ed.): Integration of Activity in the Higher Plant. Pp. 251-273. Cambridge University Press, Cambridge - London - New York - Melbourne 1977.

*3129 - McIVOR, J.G.: The effect of waterlogging on the growth of *Stylosanthes guyanensis*. - Trop. Grasslands 10: 173-178, 1976.

3130 - McMICHAEL, B.L., ELMORE, C.D.: Proline accumulation in water stressed cotton leaves. - Crop Sci. 17: 905-908, 1977.

3131 - McMICHAEL, B.L., HANNY, B.W.: Endogenous levels of abscisic acid in water-stressed cotton leaves. - Agron. J. 69: 979-982, 1977.

3132 - McPHERSON, H.G., BOYER, J.S.: Regulation of grain yield by photosynthesis in maize subjected to a water deficiency. - Agron. J. 69: 714-718, 1977.

*3133 - McPHERSON, H.G., CHU, A.C.P.: Some problems and possibilities in plant water relations research. - In: Proceedings of Soil and Plant Water Symposium. Pp. 34-42. Palmerston North 1976.

*3134 - MEDINA, E., DELGADO, M.: Photosynthesis and night $CO_2$ fixation in *Echeveria columbiana* v. Poellnitz. - Photosynthetica 10: 155-163, 1976.

3135 - MEDINA, E., DELGADO, M., TROUGHTON, J.H., MEDINA, J.D.: Physiological ecology of $CO_2$ fixation in *Bromeliaceae*. - Flora 166: 137-152, 1977.

*3136 - MEGO, V.: Vplyv vlhkosti substrátu na rast a metabolizmus základných živín lucerny siatej (*Medicago sativa* L.). 1. Dynamika rastu nadzemných orgánov. [Effect of substrate humidity on growth and metabolism of basic nutrients of *Medicago sativa* L. 1. Rate of growth of above-ground organs.] - In: Vedecké Práce Výskumného Ústavu Rastlinnej Výroby v Piešťanoch 13. Pp. 49-60. Piešťany 1976. [In Slov, ab: E, R.]

3137 - MEGO, V.: Vplyv vlhkosti substrátu na rast a metabolizmus základných živín lucerny siatej (*Medicago sativa* L.). 2. Rast a vývin v kontrolovaných podmienkách prostredia. [Effect of substrate humidity on growth and metabolism of basic nutrients of *Medicago sativa* L. 2. Growth and development under controlled conditions.] - In: Vedecké Práce Výskumného Ústavu Rastlinnej Výroby v Piešťanoch 14. Pp. 137-145. Piešťany 1977. [In Slov, ab: E, R.]

3138 - MEGO, V., ERDELSKÝ, K.: Freie Aminosäuren in der Luzerne ( *Medicago sativa* L.) - Acta Fac. Rerum Nat. Univ. Comenianae, Physiol. Plant. 13: 49-55, 1977.

3139 - MEIDNER, H.: Sap exudation via the epidermis of leaves. - J. exp. Bot. 28: 1408-1416, 1977.

3140 - MEIRI, A., KAMBUROV, J., SHALHEVET, J.: Transpiration effects on leaching fractions. - Agron. J. 69: 779-782, 1977.

3141 - MICHEL, B.E.: A model relating root permeability to flux and potentials. Application to existing data from soybean and other plants. - Plant Physiol. 60: 259-264, 1977

3142 - MICHEL, B.E.: A miniature stem thermocouple hygrometer. - Plant Physiol. 60: 645-647, 1977.

3143 - MIKHAÏLOVA, A.V.: Izmenenie soderzhaniya nukleinovykh kislot v list'yakh yachmenya v svyazi s zatopleniem i obogashcheniem vitaminom PP. [Changes in the nucleic acid content in barley leaves in connection with flooding and enrichment with vitamin PP.] - Fiziol. Biokhim. kul't. Rast. 9: 594-599, 1977. [In R, ab: E.]

3144 - MILBURN, J.A., ZIMMERMANN, M.H.: Preliminary studies on sapflow in *Cocos nucifera* L. I. Water relations and xylem transport. - New Phytol. 79: 535-541, 1977.

3145 - MILBURN, J.A., ZIMMERMANN, M.H.: Preliminary studies on sapflow in *Cocos nucifera* L. II. Phloem transport. - New Phytol. 79: 543-558, 1977.

*3146 - MILES, G.E.: Gas exchange measurement of plant leaves. - CSIRO, Aust. Div. Land Res. Manage. Tech. Paper 1: 1-7, 1976.

3147 - MILES, W.G., DAINES, R.H., RUE, J.W.: Presymptomatic egress of *Xanthomonas pruni* from infected peach leaves. - Phytopathology 67: 895-897, 1977.

3148 - MILFORD, G.F.J., CORMACK, W.F., DURRANT, M.J.: Effects of sodium chloride on
       water status and growth of sugar beet. - J. exp. Bot. 28: 1380-1388, 1977.

3149 - MILLER, D.E., BURKE, D.W.: Effect of temporary excessive wetting on soil aer-
       ation and fusarium root rot of beans. - Plant Dis. Rep. 61: 175-179, 1977.

3150 - MILLER, D.G., MANNING, C.E., TEARE, I.D.: Effects of soil water levels on com-
       ponents of growth and yield in peas. - J. Amer. Soc. hort. Sci. 102: 349-351,
       1977.

3151 - MILLER, R.D., BRESLER, E.: A quick method for estimating soil water diffusi-
       vity functions. - Soil Sci. Soc. Amer. J. 41: 1020-1022, 1977.

3152 - MINKOV, I., KIMENOV, G., KALUCHEVA, I.: Structural characteristics of the
       chloroplasts of *Haberlea rhodopensis* Friv. upon drying and restoration. -
       Dokl. Bolg. Akad. Nauk. 30: 897-900, 1977.

3153 - MITTELHEUSER, C.J.: Rapid ultrastructural recovery of water stressed leaf
       tissue. - Z. Pflanzenphysiol. 82: 458-461, 1977.

*3154 - MIX, G.P., MARSCHNER, H.: Einfluss exogener und endogener Faktoren auf den
       Calciumgehalt von Paprika- und Bohnenfrüchten. - Z. Pflanzenernähr. Bodenk.
       139: 551-563, 1976.

3155 - MOLDAU, H.: Maximization of the plant reproductive yield under water stress.
       In: UNGER, K. (ed.): Biophysikalische Analyse Pflanzlicher Systeme. Pp. 140-
       145. VEB Gustav Fischer Verlag, Jena 1977.

3156 - MOLDAU, H., KAROLIN, A.: Effect of the reserve pool on the relationship be-
       tween respiration and photosynthesis. - Photosynthetica 11: 38-47, 1977.

3157 - MOLDAU, Kh.A.: Ust'itsa - universal'nye regulyatory fotosinteza. [Stomata -
       versatile controllers of photosynthesis.] - Fiziol. Rast. 24: 969-975, 1977.
       [In R, ab: E.]

3158 - MOLZ, F.J., BROWNING, V.D.: Effect of vegetation on landfill stabilization. -
       Ground Water 15: 409-415, 1977.

3159 - MONTENEGRO, G., RIVEROS de la PUENTE, F.: Comparison of differential environ-
       mental responses of *Colliguaja odorifera*. - Flora 166: 125-135, 1977.

3160 - MOONEY, H.A., BJÖRKMAN, O., COLLATZ, G.J.: Photosynthetic acclimation to tem-
       perature and water stress in the desert shrub *Larrea divaricata*. - Carnegie Inst.
       Year Book 76: 328-335, 1977.

3161 - MOONEY, H.A., WEISSER, P.J., GULMON, S.L.: Environmental adaptations of the
       Atacaman desert cactus *Copiapoa haseltoniana*. - Flora 166: 117-124, 1977.

3162 - MOORBY, J.: Integration and regulation of translocation within the whole
       plant. - In: JENNINGS, D.H. (ed.): Integration of Activity in the Higher Plant.
       Pp. 425-454. Cambridge University Press, Cambridge - London - New York -
       Melbourne 1977.

3163 - MOORMANN, F.R., VELDKAMP, W.J., BALLAUX, J.C.: The growth of rice on a topo-
       sequence - a methodology. - Plant Soil 48: 565-580, 1977.

*3164 - MOREAU, F.: Types stomatiques et trichome foliaire des Saxifragoïdées (Saxi-
       fragacées.) - Rev. Cytol. Biol. vég. 39: 329-341, 1976.

3165 - MORESHET, S., STANHILL, G., FUCHS, M.: Effect of increasing foliage reflec-
       tance on $CO_2$ uptake and transpiration resistance of a grain sorghum crop. -
       Agron. J. 69: 246-250, 1977.

3166 - MORGAN, J.M.: Differences in osmoregulation between wheat genotypes. - Nature
       270: 234-235, 1977.

3167 - MORGAN, J.M.: Changes in diffusive conductance and water potential of wheat
       plants before and after anthesis. - Aust. J. Plant Physiol. 4: 75-86, 1977.

3168 - MORGAN, P.W., JORDAN, W.R., DAVENPORT, T.L., DURHAM, J.I.: Abscission responses
       to moisture stress, auxin transport inhibitors, and Ethephon. - Plant Physiol.
       59: 710-712, 1977.

3169 - MOROHASHI, Y., SHIMOKORIYAMA, M.: Water content and mitochondrial activities
       in imbibitional phase of germination of *Phaseolus mungo* seeds. - Z. Pflanzen-
       physiol. 82: 173-178, 1977.

*3170 - MOROT-GAUDRY, J.F., BETHENOD, O., CHARTIER, M., CHARTIER, P.: Photosynthèse
comparée d'un Maïs normal (W 64 A) et d'un Maïs mutant opaque 2 (W 64 A O$_2$). -
Physiol. vég. 14: 595-606, 1976.

3171 - MORRALL, R.A.A.: A preliminary study of the influence of water potential on
sclerotium germination in *Sclerotinia sclerotiorum*. - Can. J. Bot. 55: 8-11,
1977.

3172 - MOSS, R.: An autoradiographic technique for the location of conditioning water
in wheat at the cellular level. - J. Sci. Food Agr. 28: 23-33, 1977.

3173 - MOSSER, J.L., MOSSER, A.G., BROCK, T.D.: Photosynthesis in the snow: the alga
*Chlamydomonas nivalis (Chlorophyceae)*. - J. Phycol. 13: 22-27, 1977.

3174 - MOTOOKA, P.S., CORBIN, F.T., WORSHAM, A.D.: Uptake and translocation of Sandoz
6706 in soybean and sicklepod. - Weed Sci. 25: 30-35, 1977.

*3175 - MOURAVIEFF, I.: Effet de l'hydratation et de l'obscurite prolongee sur les
epidermes foliaires detaches. Variations du potentiel osmotique et de l'amidon
dans les cellules stomatiques. - Bull. Mensuel Soc. Linn. Lyon 45: 257-263,
1976.

3176 - MUCHENA, S.C., GROGAN, C.O.: Effects of seed size on germination of corn (*Zea
mays*) under simulated water stress conditions. - Can. J. Plant Sci. 57: 921-
923, 1977.

*3177 - MÜHLE, H.: Beziehungen zwischen Wasserangebot, Stoffproduktion und Ertrags-
bildung bei Winterweizen. - Wiss. Z. Humboldt-Univ., Berlin, math.-naturwiss.
Reihe 25: 818-823, 1976.

3178 - MURASE, H., MERVA, G.E.: Static elastic modulus of tomato epidermis as affected by water potential. - Trans. ASAE 20: 594-597, 1977.

*3179 - MUROMTSEV, N.A.: Ob ispol'zovanii termodinamicheskogo potentsiala pochvennoĭ
vlagi v issledovaniyakh po gidrofizike pochv i rastenii. [Use of thermodynamic
potential for studying soil moisture.] - Pochvovedenie 3: 42-52, 1976. [In R,
ab: E.]

3180 - MURTAGH, G.J.: The climate induced probability distribution of short-term
responses of a tropical grass to nitrogen fertilizer. - Aust. J. exp. Agr.
anim. Husb. 17: 614-620, 1977.

3181 - MUSICK, J.T., WIESE, A.F., ALLEN, R.R.: Management of bed-furrow irrigated
soil with limited and no-tillage systems. - Trans. ASAE 20: 666-672, 1977.

3182 - NAGARAJAH, S., RATNASOORIYA, G.B.: Studies with antitranspirants on tea (*Camellia sinensis* L.) - Plant Soil. 48: 185-197, 1977.

3183 - NAIRIZI, S., RYDZEWSKI, J.R.: Effects of dated soil moisture stress on crop
yields. - Exp. Agr. 13: 51-59, 1977.

3184 - NARAIN, P., SINGH, B., PAL, B.: Note on the effect of quality and depth of
irrigation on the performance of wheat grown on soil with different levels of
salinity. - Ind. J. agr. Sci. 47: 637-639, 1977.

*3185 - NASSERY, H.: The effect of salt and osmotic stress on the retention of potass-
ium by excised barley and bean roots. - New Phytol. 75: 63-67, 1975.

*3186 - NASSERY, H., DIDEHVAR, F.: The effect of cycloheximide on absorption of ions
by bean (*Phaseolus vulgaris* L.) roots. - Ann. Bot. 40: 43-48, 1976.

3187 - NAUTIYAL, D.D., SINGH, S., PANT, D.D.: Epidermal structure and ontogeny of
stomata in *Gnetum gnemon, G. montanum* and *G. ula*. - Phytomorphology 26: 282-
296, 1977.

3188 - NAVARA, J.: Príspevok k štúdiu vodného režimu marhule. [Contribution to the
study of the water balance of the apricot.] - In: HUZULÁK, J., MASAROVIČOVÁ,
E. (ed.): Fotosyntéza a Vodný Režim Drevín. Pp. 106-114. Modra-Piesky 1977.
[In Slov, ab: E.]

3189 - NAVEH, H.: Drought and limestone-tolerant shrubs and trees for reclamation of
Mediterranean landscapes. - Isr. J. Bot. 26: 50, 1977.

*3190 - NAYAR, B.K., RAI, R., VATSALA, P.: Dermal morphology of *Vanilla planifolia*
         Andr. and *V. wightii* Lindl. - Proc. Ind. Acad. Sci., Sect. B. 84: 173-179, 1976.

*3191 - NECHIPORENKO, G.A., GRINEVA, G.M.: Vliyanie raznykh srokov zatopleniya na
         raspredelenie $C^{14}$-sakharozy-U v rastenii kukuruzy. [Effect of duration of
         inundation on distribution of $^{14}C$-sucrose-U in maize.] - Fiziol. Rast. 23: 984
         -989, 1976. [In R, ab: E.]

 3192 - NEFF, E.L., WIGHT, J.R.: Overwinter soil water recharge and herbage production
         as influenced by contour furrowing on eastern Montana Rangelands. - J. Range
         Manage. 30: 193-195, 1977.

 3193 - NEJA, R.A., WILDMAN, W.E.: Vineyard irrigation in the Salinas Valley. - Calif.
         Agr. 31: 20-21, 1977.

 3194 - NEJA, R.A., WILDMAN, W.E., AYERS, R.S., KASIMATIS, A.N.: Grapevine response to
         irrigation and trellis treatments in the Salinas Valley. - Amer. J. Enol.
         Vitic. 28: 16-26, 1977.

 3195 - NELSON, C.J., DUNN, J.H., COUTTS, J.H.: Growth responses of tall fescue and
         bermudagrass to leaf application of ancymidol. - Agron. J. 69: 61-64, 1977.

 3196 - NELSON, L.R., GALLAHER, R.N., BRUCE, R.R., HOLMES, M.R.: Production of corn
         and sorghum grain in double cropping systems. - Agron. J. 69: 41-45, 1977.

 3197 - NELSON, S.D., MAYO, J.M.: Low $K^+$ in *Paphiopedilum leeanum* leaf epidermis: Im-
         plications for stomatal functioning. - Can. J. Bot. 55: 489-495, 1.977.

 3198 - NICASTRO, C.: Relazione tra condizioni climatiche e produzione di una coltura.
         Relazione tra precipitazioni e produzione del sesamo nella zona di Maracay -
         Venezuela. [Relation between climatological conditions and crop yield. Rela-
         tion between precipitation and sesame production in Maracay, Venezuela.] -
         Riv. Agr. subtrop. trop. 71: 59-67, 1977. [In Ital, ab: E.]

*3199 - NICOU, R.: Le labour, technique d'economie de l'eau pour la zone Sahelo-Sou-
         danienne ouest Africaine. - In: Efficiency of Water and Fertilizer Use in
         Semi-Arid Regions. A Technical Document. Pp. 75-99. International Atomic
         Energy Agency, Vienna 1976.

*3200 - NIELSEN, D.R.: Solute and water movement under arid and semi-arid conditions.
         - In: Efficiency of Water and Fertilizer Use in Semi-Arid Regions. A Technical
         Document. Pp. 181-193. International Atomic Energy Agency, Vienna 1976.

 3201 - NIKOLAEV, G.M., MATORIN, D.N., AKSENOV, S.I.: Sostoyanie vody i funktsioniro-
         vanie pervichnykh reaktsii fotosinteza u lishainika *Placoleconora melanophthal-
         ma* iz kholodnykh pustyn' vostochnogo Pamira. [State of water and primary pho-
         tosynthetic reactions in lichen *Placoleconora melanophthalma* from cold deserts
         of eastern Pamir.] - Nauch. Dokl. vyssh. Shkoly, biol. Nauki 20: 101-105, 1977.
         [In R.]

 3202 - NISHIDA, K.: $CO_2$ fixation in leaves of a CAM plant without lower epidermis and
         the effect of $CO_2$ on their deacidification. - Plant Cell Physiol. 18: 927-930,
         1977.

 3203 - NNYAMAH, J.U., BLACK, T.A.: Rates and patterns of water uptake in a Douglas-
         fir forest. - Soil Sci. Soc. Amer. J. 41: 972-979, 1977.

 3204 - NOBEL, P.S.: Water relations of flowering of *Agave deserti*. - Bot. Gaz. 138:
         1-6, 1977.

 3205 - NOBEL, P.S.: Water relations and photosynthesis of a barrel cactus, *Ferocactus
         acanthodes*, in the Colorado desert. - Oecologia 27: 117-133, 1977.

 3206 - NOBEL, P.S.: Internal leaf area and cellular $CO_2$ resistance: Photosynthetic
         implications of variations with growth conditions and plant species. - Physiol.
         Plant. 40: 137-144, 1977.

 3207 - NOKS, P.P., KONONENKO, A.A., RUBIN, A.B.: Nemonotonnyi kharakter sorbtsii i
         desorbtsii vody preparatami khromatoforov fotosinteziruyushchikh purpurnykh
         bakterii. [Non-monotonous character of water sorption and desorption by chro-
         matophore preparations from photosynthesizing bacteria.]- Biofizika 22: 721-
         723, 1977. [In R, ab: E.]

3208 - NOSE, A., SHIROMA, M., MIYAZATO, K., MURAYAMA, S.: Studies on matter production in pineapple plants. 1. Effects of light intensity in light period on the $CO_2$ exchange and $CO_2$ balance of pineapple plants. - Jap. J. Crop Sci. 46: 580-587, 1977.

3209 - NULSEN, R.A., THURTELL, G.W., STEVENSON, K.R.: Response of leaf water potential to pressure changes at the root surface of corn plants. - Agron. J. 69: 951-954, 1977.

3210 - NYE, P.H.: The rate-limiting step in plant nutrient absorption from soil. - Soil Sci. 123: 292-297, 1977.

*3211 - OGDEN, J.: Notes on the influence of drought on the bush remnants of the Manawatu lowlands. - Proc. N. Zeal. ecol. Soc. 23: 92-98, 1976.

*3212 - OKUNTSOV, M.M., RON'ZHINA, O.A., LIPNYAGOVA, L.A.: Deĭstvie pochvennoĭ zasukhi na sistemu adenilovykh nukleotidov yarovoĭ pschenitsy. [Effect of soil drought on system of spring wheat adenylic nucleotides.] - Fiziol. Biokhim. kul't. Rast. 8: 257-261, 1976. [In R, ab: E.]

*3213 - OLOFINBOBA, M.O., FAWOLE, M.O.: Response of six-month-old seedlings of *Theobroma cacao* to foliar treatment with Spruce Seal. - Turrialba 26: 365-370, 1976.

3214 - OPITZ von BOBERFELD, W., BOEKER, P.: Einfluss differenzierter Grundwasserstände, Nutzungsarten und Düngungsintensitäten auf die Pflanzengesellschaften des Dauergrünlandes, dargestellt am Grundwasserstandsversuch Boker Heide/Westfalen. - Z. Kulturtechnik Flurberein. 18: 13-22, 1977.

3215 - ORD, G.N.S.G., CAMERON, I.F., FENSOM, D.S.: The effect of pH and ABA on the hydraulic conductivity of *Nitella* membranes. - Can. J. Bot. 55: 1-4, 1977.

3216 - ORPHANOS, P.I.: Emergence of *Phaseolus vulgaris* seedlings from wet soil. - J. hort. Sci. 52: 447-455, 1977.

3217 - OSTER, J.D., RHOADES, J.D.: Various indices for evaluating the effective salinity and sodicity of irrigation waters. - In: DREGNE, H.E. (ed.): Managing Saline Water for Irrigation. Pp. 1-14. Texas Technical University, Lubbock 1977.

3218 - O'TOOLE, J.C., LUDFORD, P.M., OZBUN, J.L.: Gas exchange and enzyme activity during leaf expansion in *Phaseolus vulgaris* L. - New Phytol. 78: 565-571, 1977.

3219 - O'TOOLE, J.C., OZBUN, J.L., WALLACE, D.H.: Photosynthetic response to water stress in *Phaseolus vulgaris*. - Physiol. Plant. 40: 111-114, 1977.

3220 - OUTLAW, W.H., Jr., LOWRY, O.H.: Organic acid and potassium accumulation in guard cells during stomatal opening. - Proc. nat. Acad. Sci. USA 74: 4434-4438, 1977.

*3221 - PADMANI, D.R., MOTWANI, V.T., PATEL, J.C.: Note on the effect of time of irrigation and nitrogen on yield of wheat "J1-7". - GAU Res. J. 2: 65-66, 1976.

3222 - PAKIANATHAN, S.W.: Some factors affecting yield response to stimulation with 2-chloroethylphosphonic acid. - J. Rubber Res. Inst. Malaysia 25: 50-60, 1977.

*3223 - PAL, S.K., SANDHU, M.K., DAS, T.M.: Physiological studies on submerged rice. I. Effect of submergence on germination, growth, dry matter accumulation, and carbohydrate and nitrogen metabolism in rice seedlings. - Ind. Biol. 8: 27-33, 1976.

*3224 - PALEVITZ, B.A.: Ions and stomatal differentiation. - Plant Physiol 57 (Suppl.): 43, 1976.

*3225 - PALIT, P., KUNDU, A., MANDAL, R.K., SIRCAR, S.M.: Varietal and seasonal differences in growth and yield parameters of dwarf and tall varieties of rice grown with constant doses of fertilizers. - Ind. J. agr. Sci. 46: 327-337, 1976.

*3226 - PALLAS, J.E., Jr., MICHEL, B.E.: Comparison of leaf VS stem psychrometers for measuring changes in plant water potential. - Plant Physiol. 57 (Suppl.): 44, 1976.

3227 - PALLAS, J.E., Jr., STANSELL, J.R., BRUCE, R.R.: Peanut seed germination as related to soil water regime during pod development. - Agron. J. 69: 381-382, 1977.

*3228 - PALTA, J.P., LEVITT, J., STADELMANN, E.J.: Post-thawing behavior of freeze-stressed *Allium cepa* cells. - Plant Physiol. 57 (Suppl.): 34, 1976.

3229 - PALTA, J.P., LEVITT, J., STADELMANN, E.J.: Freezing injury in onion bulb cells. I. Evaluation of the conductivity method and analysis of ion and sugar efflux from injured cells. - Plant Physiol. 60: 393-397, 1977.

3230 - PALTA, J.P., LEVITT, J., STADELMANN, E.J.: Freezing injury in onion bulb cells. II. Post-thawing injury or recovery. - Plant Physiol. 60: 398-401, 1977.

3231 - PALTA, J.P., LEVITT, J., STADELMANN, E.J., BURKE, M.J.: Dehydration of onion cells: A comparison of freezing vs. desiccation and living vs. dead cells. - Physiol. Plant. 41: 273-279, 1977.

*3232 - PALTA, J.P., STADELMANN, E.J.: The effect of turgor pressure on water permeability of *Allium cepa* epidermis cell membranes. - Plant Physiol. 57 (Suppl.): 79, 1976.

3233 - PALTA, J.P., STADELMANN, E.J.: Effect of turgor pressure on water permeability of *Allium cepa* epidermis cell membranes. - J. Membrane Biol. 33: 231-247, 1977.

3234 - PALZKILL, D.A., TIBBITTS, T.W.: Evidence that root pressure flow is required for calcium transport to head leaves of cabbage. - Plant Physiol 60: 854-856, 1977.

*3235 - PANDE, H.K., RAVINDRANATH, E.: Leaf infiltration as an index of the time of irrigation of wheat field. - Ind. J. agr. Sci. 45: 539-544, 1975.

*3236 - PANKUCSI, E.: Epidermis studies on tobacco varieties. - Acta Agron. Acad. Sci. Hung. 25: 89-97, 1976.

3237 - PAPAGEORGIOU, G.C., ISAAKIDOU, J.: A comparative study of glutaraldehyde and dimethylsubermidate as protein crosslinking agents for chloroplast membranes. - In: PACKER, L., PAPAGEORGIOU, G.C., TREBST, A. (ed.): Bioenergetics of Membranes. Pp. 257-268. Elsevier/North-Holland Biomedical Press, Amsterdam - Oxford - New York 1977.

*3238 - PARMELE, L.H., JACOBY, E.L., Jr.: Estimating Evapotranspiration under Nonhomogenous Field Conditions. - Agricultural Research Service, U.S. Department of Agriculture, University Park 1975.

3239 - PARRA, M.A., ROMERO, G.C.: Salt tolerance of beans under a sand mulch culture. - In: DREGNE, H.E. (ed.): Managing Saline Water for Irrigation. Pp. 220-235. Texas Technical University, Lubbock 1977,

3240 - PARRISH, D.J., LEOPOLD, A.C.: Transient changes during soybean imbibition. - Plant Physiol. 59: 1111-1115, 1977.

3241 - PATE, J.S., SHARKEY, P.J., ATKINS, C.A.: Nutrition of a developing legume fruit. Functional economy in terms of carbon, nitrogen, water. - Plant Physiol. 59: 506-510, 1977.

*3242 - PATEL, J.C., PATEL, A.S.: Water use management as per critical stages in wheat. - GAU Res. J. 1: 47-51, 1975.

*3243 - PATEL, J.C., PATEL, A.S.: Correlation of evaporation values of different evaporimeters with consumptive use of water by wheat. - GAU Res. J. I: 11-15, 1975.

*3244 - PATEL, P.M., WALLACE, A., WALLIHAN, E.F.: Influence of salinity and N-P fertility levels on mineral content and growth of sorghum in sand culture. - Agron. J. 67: 622-625, 1975.

3245 - PATTERSON, D.T., BUNCE, J.A., ALBERTE, R.S., VAN VOLKENBURGH, E.: Photosynthesis in relation to leaf characteristics of cotton from controlled and field environments. - Plant Physiol. 59: 384-387, 1977.

3246 - PAUL, A.K., MUKHERJI, S.: Germination behaviour and metabolism of rice seeds under water-logged condition. - Z. Pflanzenphysiol. 82: 117-124, 1977.

3247 - PEACOCK, W.L., ROLSTON, D.E., ALJIBURY, F.K., RAUSCHKOLB, R.S.: Evaluating drip, flood, and sprinkler irrigation of wine grapes. - Amer. J. Enol. Vitic. 28: 193-195, 1977.

3248 - PECK, A.J., LUXMOORE, R.J., STOLZY, J.L.: Effects of spatial variability of soil hydraulic properties in water budget modeling. - Water Resour. Res. 13: 348 - 353, 1977.

3249 - PEET, M.M., BRAVO, A., WALLACE, D.H., OZBUN, J.L.: Photosynthesis, stomatal resistance, and enzyme activities in relation to yield of field-grown dry bean varieties. - Crop Sci. 17: 287-293, 1977.

3250 - PEET, M.M., OZBUN, J.L., WALLACE, D.H.: Physiological and anatomical effects of growth temperature on *Phaseolus vulgaris* L. cultivars. - J. exp. Bot. 28: 57-69, 1977.

3251 - PEGELOW, E.J., Jr., BUXTON, D.R., BRIGGS, R.E., MURAMOTO, H., GENSLER, W.G.: Canopy photosynthesis and transpiration of cotton as affected by leaf type. - Crop Sci. 17: 1-4, 1977.

3252 - PEISKER, M.: Transpiration and $CO_2$ uptake at varying stomatal aperture. - In: UNGER, K. (ed.): Biophysikalische Analyse Pflanzlicher Systeme. Pp. 151-153. VEB Gustav Fischer Verlag, Jena 1977.

3253 - PELEVINA, L.V.: Vliyanie mikroélementov margantsa i tsinka na intensivnost' transpiratsii list'ev yabloni. [Effect of manganese and zinc on apple-tree leaves transpiration rate.] - Fiziol. Biokhim. kul't. Rast. 9: 190-192, 1977. [In R, ab: E.]

3254 - PELKONEN, P., HARI, P., LUUKKANEN, O.: Decrease of $CO_2$ exchange in Scots pine after naturally occurring or artificial low temperatures. - Can. J. Forest Res. 7: 462-468, 1977.

3255 - PEMADASA, M.A.: Stomatal responses to high temperature in darkness. - Ann. Bot. 41: 969-976, 1977.

3256 - PENKA, M.: Transpirace lesních dřevin. [Transpiration in forest species.] - In: Dni Rastlinnej Fyziológie. Pp. 168-171, 308. Slovenská Botanická Spoločnost SAV, Bratislava 1977. [In Czech, ab: R, E.]

3257 - PENKA, M.: Vodní provoz dřevin. [The water regime of woody species.] - In: HUZULÁK, J., MASAROVIČOVÁ, E. (ed.): Fotosyntéza a Vodný Režim Drevín. Pp. 3-46. Modra-Piesky 1977. [In Czech, ab: E.]

3258 - PENNING de VRIES, F.W.T.: Simulation der Assimilation und Transpiration der Pflanzendecke nach grundlegenden Gesetzen. - In: UNGER, K. (ed.): Biophysikalische Analyse Pflanzlicher Systeme. Pp. 107-114. VEB Gustav Fischer Verlag, Jena 1977.

3259 - PENNING de VRIES, F.W.T., Van LAAR, H.H.: Substrate ulilization in germinating seeds. - In: LANDSBERG, J.J., CUTTING, C.V. (ed.): Environmental Effects on Crop Physiology. Pp. 217-228. Academic Press, London - New York - San Francisco 1977.

*3260 - PEREIRA, J.S., KOZLOWSKI, T.T.: Diurnal and seasonal changes in water balance of *Abies balsamea* and *Pinus resinosa*. - Oecol. Plant. 11: 397-412, 1976.

3261 - PEREIRA, J.S., KOZLOWSKI, T.T.: Water relations and drought resistance of young *Pinus banksiana* and *P. resinosa* plantation trees. - Can. J. Forest Res. 7: 132-137, 1977.

3262 - PEREIRA, J.S., KOZLOWSKI, T.T.: Influence of light intensity, temperature, and leaf area on stomatal aperture and water potential of woody plants. - Can. J. Forest Res. 7: 145-153, 1977.

3263 - PEREIRA, J.S., KOZLOWSKI, T.T.: Variations among woody angiosperms in response to flooding. - Physiol. Plant. 41: 184-192, 1977.

3264 - PERESYPKIN, V.F., DRAPATYĬ, N.A.: Osobennosti nekotorykh fiziologo-biokhimicheskikh protsessov v list'yakh yachmenya pri porazhenii ikh gribom *Rhynchosporium graminicola* Heins. [Peculiarities of certain physiological and biochemical processes in barley leaves with their affection by *Rhynchosporium graminicola* Heins. fungus.] - Fiziol. Biokhim. kul't. Rast. 9: 377-380, 1977. [In R.]

3265 - PEROSIAN, G.P.: Chemical reclamation of solonetz-solonchaks in the Ararat Plain
       and the possibility of utilizing saline waters for leaching and irrigation. -
       In: DREGNE, H.E. (ed.): Managing Saline Water for Irrigation. Pp. 466-479.
       Texas Technical University, Lubbock 1977.

*3266 - PERRIER, A.: Methods of observation of heat and mass transfer in the lower
       atmosphere and in plant canopies. - In: De VRIES, D.A., AFGAN, N.H. (ed.):
       Heat and Mass Transfer in the Biosphere. Part 1. Transfer Processes in the
       Plant Environment. Pp. 229-249. Scripta Book Company, Washington 1975.

3267 - PERRY, L.J., Jr., COMPTON, W.A.: Serial measures of dry matter accumulation
       and forage quality of leaves, stalks, and ears of three corn hybrids. - Agron.
       J. 69: 751-755, 1977.

*3268 - PERSAUD, N., LOCASCIO, S.J., GERALDSON, C.M.: Influence of fertilizer rate and
       placement and irrigation method on plant nutrient status, soil soluble salt
       and root distribution of mulched tomatoes. - Soil Crop Sci. Soc. Florida Proc.
       36: 121-125, 1976.

3269 - PETERSON, H.B., SHUPE, J.L.: Problems of managing geothermal waters for irri-
       gation. - In: DREGNE, H.E. (ed.): Managing Saline Water for Irrigation. Pp.
       211-219. Texas Technical University, Lubbock 1977.

3270 - PETINOV, N.S., IVANOV, V.P., KIRILLINA,V.I., GOLOVATYĬ, V.G., KALIMULINA, Kh.K:
       Dinamika nakopleniya sukhogo veshchstva i plasticheskikh soedineniĭ v raste-
       niyakh yachmenya v zavisimosti ot vlazhnosti pochvy i urovnya mineral'nogo
       pitaniya. [Dynamics of accumulation of dry matter and plastic compounds in
       barley plants as a function of soil humidity and level of mineral nutrition.]
       - Fiziol Rast. 24: 593-600, 1977. [In R, ab: E.]

*3271 - PETKOV, P.S.: Posledeĭstvie na toreneto v"rkhu posevnite kachestva na semenata
       i dobiva pri fasula. [Aftereffect of fertilization upon the seeding qualities
       and the yield of beans.] - Rasteniev"dni Nauki 12: 93-99, 1975. [In Bulg,
       ab: R, E.]

*3272 - PETKOV, P.S.: Vliyanie na usloviyata na otglezhdane v"rkhu biologichnite oso-
       benosti na nyakoi sortove fasul. [Growing conditions as affecting the biologi-
       cal peculiarities of some bean varieties.] - Rasteniev"dni Nauki 12: 45-55,
       1975. [In Bulg, ab: R, E.]

3273 - PETRASOVITS, I.: Ökológiai vizsgálatok öntözött növényállományokban. [Ecologi-
       cal investigations in irrigated plant stands.] - Növénytermelés 26: 59-67,
       1977. [In Hung., ab: E.]

*3274 - PETROV, A.P.: O pogloshenii vodyanykh parov rasteniyami. [On the absorption
       of water vapour by plants.] - Bot. Zh. 60: 309-310, 1975. [In R.]

*3275 - PETROVA, V.N.: Abstsizovaya kislota - gormon rasteniĭ. [Abscisic acid - the
       hormone of plants.] - Bot. Zh. 61: 1004-1016, 1976. [In R.]

3276 - PEVELING, E.: Die Ultrastruktur einiger Flechten nach langen Trockenzeiten. -
       Protoplasma 92: 129-136, 1977.

3277 - PFENDER, W.F., HINE, R.B., STANGHELLINI, M.E.: Production of sporangia and
       release of zoospores by *Phytophthora megasperma.* in soil. - Phytopathology 67:
       657-663, 1977.

3278 - PHAM THI, A.T.,VIEIRA da SILVA,J.B.: Action des déficits hydriques sur la
       photosynthèse et sur la respiration des feuilles du cotonnier. - In: MOYSE, A.
       (ed.): Les Processus de la Production Végétale Primaire. Pp. 183-202.
       Gauthier-Villars, Paris 1977.

*3279 - PHILLIPS, R.D., JENNINGS, D.H.: Succulence, cations and organic acids in
       leaves of *Kalanchoe daigremontiana* grown in long and short days in soil and
       water culture. - New Phytol. 77: 599-611, 1976.

3280 - PHIPPS, P.M., BEUTE, M.K.: Influence of soil temperature and moisture on the
       severity of *Cylindrocladium* black rot in peanut. - Phytopathology 67: 1104-
       1107, 1977.

*3281 - PIERI, C.: L'utilisation des engrais dans les sols de la zone semi-arid du
       Senegal. - In: Efficiency of Water and Fertilizer Use in Semi-Arid Regions.
       A Technical Document. Pp. 159-180. International Atomic Energy Agency, Vienna
       1976.

3282 - PILL, W.G., LANBETH, V.N.: Effects of $NH_4$ and $NO_3$ nutrition with and without pH adjustment on tomato growth, ion composition, and water relations. - J. Amer. Soc. hort. Sci. 102: 78-81, 1977.

3283 - PITMAN, M.G.: Ion transport into the xylem. - Annu. Rev. Plant Physiol. 28: 71 - 88, 1977.

3284 - PITMAN, M.G., CRAM, W.J.: Regulation of ion content in whole plants. - In: JENNINGS, D.H. (ed.): Integration of Activity in the Higher Plant. Pp. 391-424. Cambridge University Press, Cambridge - London - New York - Melbourne 1977.

3285 - PITT, J.I., HOCKING, A.D.: Influence of solute and hydrogen ion concentration on water relations of some xerophilic fungi. - J. gen. Microbiol. 101: 35-40, 1977.

3286 - PLAMENAC, N., VITKOVIĆ, M., MATIĆ, M.: Calculation and prediction of the salt regime of lowland soils under influence of groundwater salinization. - In: DREGNE, H.E. (ed.): Managing Saline Water for Irrigation. Pp. 277-289. Texas Technical University, Lubbock 1977.

3287 - PLAUT, Z., ZIESLIN, N.: The effect of canopy wetting on plant water status, $CO_2$ fixation, ion content and growth rate of "Baccara" roses. - Physiol. Plant. 39: 317-322, 1977.

*3288 - POLERECKÝ, O.: Úroda a jej komponenty u novošľachtených úrodných hybridov kukurice na zrno. [Yield and its components in new-bred hight-yielding maize hybrids.] - Rost. Výroba 22: 1021-1027, 1976. [In Slov, ab: E, R.]

3289 - POLLARD, D.F.W., LOGAN, K.T.: The effects of light intensity, photoperiod, soil moisture potential, and temperature on bud morphogenesis in *Picea* species. - Can. J. Forest Res. 7: 415-421, 1977.

*3290 - PORTAS, C.A.M., TAYLOR, H.M.: Growth and survival of young plant roots in dry soil. - Soil Sci. 121: 170-175, 1976.

3291 - POSPISHILOVA, Ya., ZIMA, Ya., SHESTAK, Z., SOLAROVA, Ya.: Vliyanie abstsisovoĭ kisloty na aktivnost' fotosistemy 2 u izolirovannykh khloroplastov. [Effect of abscisic acid on photosystem 2 activity of isolated chloroplasts.]- In: Pigment Protein Complexes in Photosynthesis. Biological Research Center, Szeged 1977. [In R.]

3292 - POSPÍŠILOVÁ, J.: Water relations in primary leaves of bean plants treated with polyethylene glycol solutions. - Biol. Plant. 19: 316-319, 1977.

3293 - POSPÍŠILOVÁ, J., ČATSKÝ, J., SOLÁROVÁ, J., ŠESTÁK, Z., TICHÁ, I., VÁCLAVÍK, J., ZIMA, J.: Limitace fotosyntézy hydratační úrovní listového mezofylu. [Limitation of photosynthesis by hydration level of leaf mesophyll.]- In: Dni Rastlinnej Fyziológie. Pp. 122-125, 297. Slovenská Botanická Spoločnost SAV, Bratislava 1977. [In Czech, ab: R. E.]

3294 - POWELL, D.B.B., THORPE, M.R.: Dynamic aspects of plant-water relations. - In: LANDSBERG, J.J., CUTTING, C.V. (ed.): Environmental Effects on Crop Physiology. Pp. 259-279. Academic Press, London - New York - San Francisco 1977.

3295 - PRATT, P.F., DAVIS, S., ADRIANO, D.C., BISHOP, S.E., LAAG, A.E.: Fate of salts from water and manure in a 4-year field experiment. - In: DREGNE, H.E. (ed.): Managing Saline Water for Irrigation. Pp. 264-276. Texas Technical University, Lubbock 1977.

3296 - PRAŽÁK, M.: Výzkum vodního režimu u jabloní v půdních lysimetrech. [Investigations of the water regime of apple trees by soil lysimeters.]- In: HUZULÁK, J., MASAROVICOVA, E. (ed.): Fotosyntéza a Vodný Režim Drevín. Pp. 123-129. Modra-Piesky 1977. [In Czech, ab: E, R.]

3297 - PREMECZ, G., OLÁH, T., GULYÁS, A., NYITRAI, Á., PÁLFI, G., FARKAS, G.L.: Is the increase in ribonuclease level in isolated tobacco protoplasts due to osmotic stress? - Plant Sci. Lett. 9: 195-200, 1977.

*3298 - PRESSLAND, A.J., BATIANOFF, G.N.: Soil water conservation under cultivated fallows in clay soils of south-western Queensland. - Aust. J. exp. Agr. anim. Husb. 16: 564-569, 1976.

3299 - PRESSMAN, E., NEGBI, M., SACHS, M., JACOBSEN, J.V.: Varietal differences in
light requirements for germination of celery (*Apium graveolens* L.) seeds and
the effects of thermal and solute stress. - Aust. J. Plant Physiol. 4: 821-
831, 1977.

3300 - PRIEHRADNÝ, S.: Response to fungus pathogen in susceptible and resistant barley cultivars. II. Water uptake. - Phytopathol. Z. 90: 1-11, 1977.

3301 - PRIEHRADNÝ, S.: Changes in the ratio of water uptake to water loss in light
and darkness followed in barley infected by the powdery mildew. - Biológia
(Bratislava) 32: 739-746, 1977.

3302 - PRIEHRADNÝ, S.: Príjem vody múčnatkou infikovaného jačmeňa rezistentného kul-
tivaru. [Water uptake by resistant cultivar of barley infected by powdery mildew.] - Biológia (Bratislava) 32: 43-51, 1977. [In Slov, ab: R, E.]

3303 - PRIEHRADNÝ, S.: Zmeny vodnej bilancie na svetle a tme pri jačmeni infikovanom
múčnatkou. [Alterations of water balance in the light and in the dark by barley infected with powdery mildew.]-In: Dni Rastlinnej Fyziológie. Pp. 176-179,
310. Slovenská Botanická Spoločnost SAV, Bratislava 1977. [In Slov, ab: E, R.]

3304 - PRIHAR, S.S., SANDHU, B.S., KHERA, K.L.: Potentialities of increasing maize
and sugarcane yields with straw mulching in Punjab. - Ind. J. Ecol. 4: 167-
176, 1977.

3305 - PRIHAR, S.S., KHERA, K.L., BAJWA, M.S.: Growth, water use and nutrient uptake
by dryland wheat, as affected by the placement of nitrogen and phosphorus. -
Ind. J. Ecol. 4: 23-31, 1977.

3306 - PROEBSTING, E.L., MIDDLETON, J.E., ROBERTS, S,: Altered fruiting and growth
characteristics of "Delicious" apple associated with irrigation method. -
HortScience 12: 349-350, 1977.

3307 - PROKHORCHIK, R.A.: Vliyanie abstsizovoǐ kisloty na fotokhimicheskuyu aktivnost'
khloroplastov lyupina. [Effect of abscisic acid on photochemical activity of
lupine chloroplasts.] - Dokl. Akad. Nauk Beloruss.SSR 21: 644-646, 1977. [In R.]

3308 - PTÁČKOVÁ, M.: Spotřeba vody u vojtěšky. [Water consumption in alfalfa.]- Rost.
Výroba 23: 185-192, 1977. [In Czech, ab: E, R, G.]

3309 - PTÁČKOVÁ, M.: Počet a velikost průduchů u vojtěšky. [Number and size of stomata in alfalfa.]-Rost. Výroba 23: 1107-1114, 1977. [In Czech, ab: E, R, G.]

3310 - PTÁČKOVÁ, M.: Vodní provoz rostlin ve vztahu k minerální výživě. [Water-rela-
tion of plants referring to mineral nutrition.] - Věstník Československé
Akad. Zeměd. 24: 142-144, 1977. [In Czech ]

3311 - PUJOL, B., VACQUÉ, A., CABELGUENNE, M., DECAU, J.: Contribution à l'étude des
relations entre la consommation d'eau et la production dans des cultures de
Maïs (*Zea Mays* L.) portant ou non le gène "opaque$_2$". Rôle du taux de fructifi-
cation. - C.R. Acad. Sci. Paris, Sér. D 284: 807-810, 1977.

3312 - PURI, J., TAL, M.: Abnormal stomatal behavior and hormonal imbalance in *flacca*,
a wilty mutant of tomato. IV. Effect of abscisic acid and water content on
RNase activity and RNA. - Plant Physiol. 59: 173-177, 1977.

3313 - QUARRIE, S.A., JONES, H.G.: Effects of abscisic acid and water stress on development and morphology of wheat. - J. exp. Bot. 28: 192-203, 1977.

3314 - RAATS, P.A.C.: Convective transport of solutes in and below the root zone. -
In: DREGNE, H.E. (ed.): Managing Saline Water for Irrigation. Pp. 290-298.
Texas Technical University, Lubbock 1977.

3315 - RACUSEN, R.H., KINNERSLEY, A.M., GALSTON, A.W.: Osmotically induced changes in
electrical properties of plant protoplast membranes. - Science 198: 405-406,
1977.

3316 - RADOMSKI, C., MADANY, R., NOZYŃSKI, A.: Effect of the slope microclimate on
yield of cereals. - Agr. Meteorol. 18: 203-209, 1977.

3317 - RAGHAVENDRA, A.S., DAS, V.S.R.: Antitranspirant activity of inhibitors of cyc-
lic photophosphorylation. - J. exp. Bot. 28: 480-483, 1977.

*3318 - RAGHU, J.S., SHARMA, S.R., CHOUBEY, S.D.: Response of different rice varieties to various moisture regimes. - JNKVV Res. J. 9: 161-162, 1975.

3319 - RAGUSE, C.A., YOUNG, J.A., EVANS, R.A.: Germination of California annual range plants in response to summer rain. - Agron. J. 69: 327-329, 1977.

3320 - RAJAGOPAL, V., BALASUBRAMANIAN, V., SINHA, S.K.: Diurnal fluctuation in relative water content, nitrate reductase and proline content in water-stressed and non-stressed wheat. - Physiol. Plant. 40: 69-71, 1977.

3321 - RAJENDRA, B.R., MUJEEB, K.A., BATES, L.S.: A modified technique to study leaf epidermis in *Triticeae*. - Stain Technol. 52: 9-12, 1977.

3322 - RAJU, V.S., RAO, P.N.: Variation in the structure and development of foliar stomata in the *Euphorbiaceae*. - Bot. J. Linn. Soc. 75: 69-97, 1977.

3323 - RAND, R.H.: Gaseous diffusion in the leaf interior. - Trans. ASAE 20: 701-704, 1977.

*3324 - RAO, P.N., RAJU, V.S.: Little known features in the foliar epidermology of some *Euphorbiaceae*. - Curr. Sci. 44: 750-752, 1975.

3325 - RAO, S.R.S., RAMAYYA, N.: Stomatogenesis in the genus *Hibiscus* L. (*Malvaceae*). - Bot. J. Linn. Soc. 74: 47-56, 1977.

*3326 - RAO, V.R., RAMACHANDRAM, M., RAO, M.S.R.M.: Plant density and geometry in relation to varietal differences and seasonal variations in rainfall for increasing and stabilizing production levels of winter sorghum in drylands. - Ind. J. agr. Sci. 46: 559-566, 1976.

3327 - RASCHKE, K.: The stomatal turgor mechanism and its responses to $CO_2$ and abscisic acid: Observations and a hypothesis. - In: MARRÈ, E., CIFERRI, O. (ed.): Regulation of Cell Membrane Activities in Plants. Pp. 173-183. Elsevier/ Nort Holland Biomedical Press, Amsterdam - Oxford - New York 1977.

3328 - RASCHKE, K.: The osmotic motor of stomatal movement. - In: JUNGREIS, A.M., HODGES, T.K., KLEINZELLER, A., SCHULTZ, S.G. (ed.): Water Relations in Membrane Transport in Plants and Animals. Pp. 47-53. Academic Press, New York - San Francisco - London 1977.

3329 - RASCHKE, K., DITTRICH, P.: [$^{14}$C]Carbon-dioxide fixation by isolated leaf epidermes with stomata closed or open. - Planta 134: 69-75, 1977.

3330 - RATHAIAH, Y.: Spore germination and mode of cotton infection by *Ramularia areola*. - Phytopathology 67: 351-357, 1977.

3331 - RATHAIAH, Y.: Stomatal tropism of *Cercospora beticola* in sugarbeet. - Phytopathology 67: 358-362, 1977.

*3332 - RATHNAM, C.K.M., RAGHAVENDRA, A.S., DAS, V.S.R.: Diversity in the arrangements of the mesophyll cells among leaves of certain $C_4$ dicotyledons in relation to $C_4$ physiology. - Z. Pflanzenphysiol. 77: 283-291, 1976.

*3333 - RATNAM, B.P., HOSMANI, M.M., JOSHI, S.N.: Moisture adequacy in relation to cropping pattern in Karnataka. - Geobios 3: 174-175, 1976.

3334 - RAVEN, J.A.: Regulation of solute transport at the non-photosynthetic cells of higher plants. - In: JENNINGS, D.H. (ed.): Integration of Activity in the Higher Plant. Pp. 73-99. Cambridge University Press, Cambridge - London - New York - Melbourne 1977.

3335 - RAWSON, H.M., BAGGA, A.K., BREMNER, P.M.: Aspects of adaptation by wheat and barley to soil moisture deficits. - Aust. J. Plant Physiol. 4: 389-401, 1977.

3336 - RAWSON, H.M., BEGG, J.E., WOODWARD, R.G.: The effect of atmospheric humidity on photosynthesis, transpiration and water use efficiency of leaves of several plant species. - Planta 134: 5-10, 1977.

*3337 - REESE, R.L., KOO, R.C.J.: Influence of fertility and irrigation on yield and leaf and soil analyses of "Temple" orange. - Proc. Florida State hort. Soc. 89: 46-48, 1976.

3338 - REESE, R.L., KOO, R.C.J.: Fertility and irrigation effects on "Temple" orange. 1. Yield and leaf analyses. - J. Amer. Soc. hort. Sci. 102: 148-151, 1977.

3339 - REGINATO, R.J., VEDDER, J.F., IDSO, S.B., JACKSON, R.D., BLANCHARD, M.B., GOETTELMAN, R.: An evaluation of total solar reflectance and spectral band ratioing techniques for estimating soil water content. - J. geophys. Res. 82: 2101-2104, 1977.

*3340 - REHM, G.W., WIESE, R.A.: Effect of method of nitrogen application on corn (*Zea mays* L.) growth on irrigated sandy soils. - Soil Sci. Soc. Amer. Proc. 39: 1217-1220, 1975.

3341 - REICOSKY, D.C., DOTY, C.W., CAMPBELL, R.B.: Evapotranspiration and soil water movement beneath the root zone of irrigated and nonirrigated millet (*Panicum miliaceum*). - Soil Sci. 124: 95-101, 1977.

3342 - REICOSKY, D.C., PETERS, D.B.: A portable chamber for rapid evapotranspiration measurements on field plots. - Agron. J. 69: 729-732, 1977.

3343 - REID, C.P.P., MEXAL, J.G.: Water stress effects on root exudation by lodgepole pine. - Soil Biol. Biochem. 9: 417-421, 1977.

3344 - RICHARDS, D.: Root-shoot interactions: A functional equilibrium for water uptake in peach (*Prunus persica* (L.) Batsch.). - Ann. Bot. 41: 279-281, 1977.

3345 - RICHTER, H.: Saugspannung und Wasserbilanz: ein Beitrag zu ihrer Interpretation. - Phyton 18: 29-41, 1977.

3346 - RIEKELS, J.W.: Nitrogen-water relationships of onions grown on organic soil. - J. Amer. Soc. hort. Sci. 102: 139-142, 1977.

3347 - RIES, S.K., WERT, V.: Growth responses of rice seedlings to triacontanol in light and dark. - Planta 135: 77-82, 1977.

*3348 - RIJKS, D.A.: Water use by irrigated cotton in the Sudan. IV. Water use, potential evaporation and yield. - J. appl. Ecol. 13: 491-506, 1976.

*3349 - RIPLEY, E.A., REDMANN, R.E.: Grassland. - In: MONTEITH, J.L. (ed.): Vegetation and the Atmosphere. Vol. 2. Case Studies. Pp. 349-398. Academic Press, London - New York - San Francisco 1975.

3350 - ROARK, B., QUISENBERRY, J.E.: Environmental and genetic components of stomatal behavior in two genotypes of upland cotton. - Plant Physiol. 59: 354-356, 1977.

3351 - ROARK, B., QUISENBERRY, J.E.: Evaluation of cotton germplasm for drought resistence. - In: Proceedings of Beltwide Cotton Production Research Conferences. Pp. 49-50. National Cotton Council, Memphis 1977.

3352 - ROBB, J., BUSCH, L., BRISSON, J.D., LU, B.C.: Ultrastructure of wilt syndrome caused by *Verticillium dahliae*. III. Chronological symptom development in sunflower leaves. - Can. J. Bot. 55: 139-152, 1977.

3353 - ROBERTS, B.R., SCHREIBER, L.R.: Influence of Dutch elm disease on resistance to water flow through roots of American elm. - Phytopathology 67: 56-59, 1977.

3354 - ROBERTS, J.: The use of tree-cutting techniques in the study of the water relations of mature *Pinus sylvestris* L. I. The technique and survey of the results. - J. exp. Bot. 28: 751-767, 1977.

3355 - ROBERTS, J., FOURT, D.F.: A small pressure chamber for use with plant leaves of small size. - Plant Soil 48: 545-546, 1977.

3356 - ROBERTS, S.W., KNOERR, K.R.: Components of water potential estimated from xylem pressure measurements in five tree species. - Oecologia 28: 191-202, 1977.

3357 - ROBINSON, F.E.: Predicted and actual yield decline from fifty percent increase salinity of the Colorado River. - In: DREGNE, H.E. (ed.): Managing Saline Water for Irrigation. Pp. 170-174. Texas Technical University, Lubbock 1977.

3358 - RODGERS, G.A.: Nitrogenase activity in *Nostoc muscorum*; recovery from desiccation. - Plant Soil 46: 671-674, 1977.

3359 - RODIONOVA, M.A., KHOLODENKO, N.Ya., GRINEVA, G.M.: Aktivnost' adenilatkinazy u rasteniĭ kukuruzy v usloviyakh pereuvlazhneniya. [The activity of adenylate kinase in corn plants under conditions of overwatering.] - Fiziol. Rast. 24: 797-802, 1977. [In R, ab: E.]

3360 - RODNEY, D.R., ROTH, R.L., GARDNER, B.R.: Citrus responses to irrigation methods.
       - Proc. Int. Soc. Citriculture. 1: 106-110, 1977.

*3361 - ROGERS, J.S., BARTHOLIC, J.F.: Estimated evapotranspiration and irrigation re-
       quirements for citrus. - Soil Crop Sci. Soc. Florida Proc. 35: 111-117, 1975.

3362 - ROGERSON, N.E., MATTHEWS, S.: Respiratory and carbohydrate changes in develop-
       ing pea (*Pisum sativum* L.) seeds in relation to their ability to withstand
       desiccation. - J. exp. Bot. 28: 304-313, 1977.

3363 - ROGOWSKI, A.S., JACOBY, E.L., Jr.: Assesment of water loss patterns with micro-
       lysimeters. - Agron. J. 69: 419-424, 1977.

*3364 - ROOK, D.A., HOBBS, J.F.F.: Soil temperatures and growth of rooted cuttings of
       radiata pine. - N. Zeal. J. Forest Sci. 5: 296-305, 1976.

3365 - ROSENTHAL, W.D., KANEMASU, E.T., RANEY, R.J., STONE, L.R.: Evaluation of an
       evapotranspiration model for corn. - Agron. J. 69: 461-464, 1977.

*3366 - ROSS, J.P.: Effect of overhead irrigation and benomyl sprays on late-season
       foliar diseases, seed infection, and yields of soybean. - Plant Dis. Rep. 59:
       809-813, 1975.

3367 - ROTH, D., TEICHARD, R., HENKEL, A.: Ergebnisse aus dem Beregnungseinsatz in
       den Trockenjahren 1975 und 1976 sowie Richtwerte und Empfehlungen zur effekti-
       ven Gestaltung und Nutzung der Beregnung. - Feldwirtschaft 5: 218-221, 1977.

3368 - ROTTINK, B.A.: Effects of a silicone antitranspirant on tree growth. - Forest
       Sci. 23: 361-362, 1977.

3369 - ROUHANI, I., KHOSH-KHUI, M.: Variations in photosynthetic rates of fourteen
       *Coleus* cultivars. - Plant Physiol. 59: 114-115, 1977.

3370 - ROUSE, G.D.: Some effects of rainfall on the growth of Corsican pine at
       Rendlesham forest, Suffolk. - Quart. J. Forest. 71: 77-80, 1977.

3371 - ROWLAND, G.G., GUSTA, L.V.: Effects of soaking, seed moisture content, tempera-
       ture and seed leakage on germination of faba beans (*Vicia faba*) and peas (*Pi-
       sum sativum*). - Can. J. Plant Sci. 57: 401-406, 1977.

3372 - RUBINSTEIN, B.: Osmotic shock inhibits auxin-stimulated acidification and
       growth. - Plant Physiol. 59: 369-371, 1977.

3373 - RUDAKOV, V.E.: O postoyanstve sootnosheniya mezhdu velichinami transpiratsii i
       massy khvoĭ u sosny obyknovennoĭ. [Constant correlation between the values of
       transpiration and mass of needles in *Pinus sylvestris*.] - Fiziol. Rast. 24:
       854-855, 1977. [In R.]

3374 - RUDICH, J., KALMAR, D., GEIZENBERG, C., HAREL, S.: Low water tensions in de-
       fined growth stages of processing tomato plants and their effects on yield and
       quality. - J. hort. Sci. 52: 391-399, 1977.

3375 - RUFFING, B.J., JENSEN, M.E., WESTERMANN, D.T.: Managing irrigation and nitrogen
       for moravian barley in Southern Idaho. - Idaho agr. exp. Sta. Curr. Inform.
       Ser. 365: 1-3, 1977.

3376 - RUNDEL, P.W., STECKER, R.E.: Morphological adaptations of tracheid structure
       to water stress gradients in the crown of *Sequoiadendron giganteum*. - Oeco-
       logia 27: 135-139, 1977.

3377 - RUNECKLES, V.C., ROSEN, P.M.: Effects of ambient ozone pretreatment on transpi-
       ration and susceptibility to ozone injury. - Can. J. Bot. 55: 193-197, 1977.

3378 - RUTTER, A.J., MORTON, A.J.: A predictive model of rainfall interception in
       forests. III. Sensitivity of the model to stand parameters and meteorological
       variables. - J. appl. Ecol. 14: 567-588, 1977.

*3379 - RYCHNOVSKÁ, M.: Transpiration in wet meadows and some other types of grass-
       land. - Folia Geobot. Phytotax. 11: 427-432, 1976.

*3380 - RYCHNOVSKÁ, M.: Alluvial grassland hydrosere: Primary production and plant
       processes. - Pol. ecol. Stud. 2: 103-112, 1976.

3381 - RYPÁČKOVÁ, M.: Vliv dřeva dekomponovaného hnědou hnilobou na odolnost mladých
       semenáčků smrku vůči suchu. [The effects of wood decomposed by brown rot upon
       the dry resistance of young spruce seedlings.] - In: HUZULÁK, J., MASAROVIČOVÁ,
       E. (ed.): Fotosyntéza a Vodný Režim Drevín. Pp. 92-98. Modra-Piesky 1977. [In
       Czech, ab: E, R.]

3382 - SAFFIGNA, P.G., KEENEY, D.R., TANNER, C.B.: Nitrogen, chloride, and water bal-
       ance with irrigated Russet Burbank potatoes in a sandy soil. - Agron. J. 69:
       251-257, 1977.

3383 - SAITO, T.: [Studies on the fruiting of muskmelons. Effects of fertilizer kind
       and level, and water supply, on quality, particularly on cracking.] - Bull.
       Coll. Agr. vet. Med., Nihon Univ. 34: 61-72, 1977. [In Jap, ab: E.]

3384 - SAJWAN, V.S., HARJAL, N., PALIWAL, G.S.: Developmental anatomy of the leaf of
       *Ficus religiosa* L. - Ann. Bot. 41: 293-302, 1977.

3385 - SALATENKO, V.N.: Fotosintez i produktivnost' kleshcheviny pri oroshenii. [Pho-
       tosynthesis and productivity of castor-oil plant with irrigation.] - Fiziol.
       Biokhim. kul't. Rast. 9: 424-431, 1977. [In R, ab: E.]

3386 - SAMIEV, Kh.S., MARFINA, K.G.: Aktivnost' RNKazy khloroplastov khlopchatnika
       pri vodnom defitsite. [Activity of chloroplast RNase in cotton under drought.]
       - Biokhimiya 42: 1361-1365, 1977. [In R, ab: E.]

3387 - SANCHEZ, S.M.: The fine structure of the guard cells of *Helianthus annuus*. -
       Amer. J. Bot. 64: 814-824, 1977.

3388 - SANDHU, B.S., HORTON, M.L.: Response of oats to water deficit. I. Physiologi-
       cal characteristics. - Agron. J. 69: 357-360, 1977.

3389 - SANDHU, B.S., HORTON, M.L.: Response of oats to water deficit. II. Growth and
       yield characteristics. - Agron. J. 69: 361-364, 1977.

*3390 - SANDHU, K.S., NIJJAR, G.S., PRIHAR, S.S.: Scheduling of irrigation to maize
       based on the induced early willing. - Ind. J. Ecol. 3: 132-140, 1976.

*3391 - SANDS, R., CLARKE, A.R.P.: Response of radiata pine to salt stress. I. Water
       relations, osmotic adjustment and salt uptake. - Aust. J. Plant Physiol. 4:
       637-646, 1977.

3392 - SAN JOSÉ, J.J.: Gas exchange in *Paspalum repens* Berg. - Rev. Brasil. Biol. 37:
       525-533, 1977.

3393 - SAN JOSÉ, J.J.: Potencial hidrico e intercambio gaseoso de *Curatella americana*
       L. en la temporada seca de la sabana Trachypogon. [Water potential and gaseous
       exchange in *Curatella americana* during dry season in Trachypogon plain.]- Acta
       Cient. Venez. 28: 373-379, 1977. [In Span, ab: E.]

*3394 - SAN JOSÉ, J.J., MEDINA, E.: Effect of fire on organic matter production and
       water balance in a tropical savana. - In: GOLLEY, F.B., MEDINA, E. (ed.):
       Tropical Ecological Systems. Pp. 251-264. Springer-Verlag, Berlin - Heidelberg
       - New York 1975.

*3395 - SAN JOSE, J.J., MEDINA, Y.E.: Organic matter production in the Trachypogon
       savana at Calabozo, Venezuela. - Trop. Ecol. 17: 113-124, 1976.

3396 - SATOH, M., KRIEDEMANN, P.E., LOVEYS, B.R.: Changes in photosynthetic activity
       and related processes following decapitation in mulberry trees. - Physiol.
       Plant 41: 203-210, 1977.

3397 - SAUGIER, B.: Micrometeorology on crops and grasslands. - In: LANDSBERG, J.J.,
       CUTTING, C.V. (ed.): Environmental Effect on Crop Physiology. Pp. 39-55. Aca-
       demic Press, London - New York - San Francisco 1977.

*3398 - SAWHNEY, J.S., MANAWI, M.: Water uptake and germination of wheat grains in
       different solutions. - Libyan J. Agr. 4: 61-64, 1975.

3399 - SCHAAD, N.W., BRENNER, D.: A bacterial wilt and root rot of sweet potato caused
       by *Erwinia chrysanthemi*. - Phytopathology 67: 302-308, 1977.

3400 - SCHEIRER, D.C., GOLDKLANG, I.J.: Pathway of water movement in hydroids of *Po-
       lytrichum commune* Hedw.(*Bryopsida*). - Amer. J. Bot. 64: 1046-1047, 1977.

3401 - SCHENK, H.E.A.: Zur osmotischen Beeinflussung der Photosynthese von *Cyanophora paradoxa* Korsch. durch Saccharose. - Arch. Microbiol. 114: 261-266, 1977.

3402 - SCHIMPF, D.J.: Seed weight of *Amaranthus retroflexus* in relation to moisture and length of growing season. - Ecology 58: 450-453, 1977.

3403 - SCHLESINGER, W.H., CHABOT, B.F.: The use of water and minerals by evergreen and deciduous shrubs in Okefenokee Swamp. - Bot. Gaz. 138: 490-497, 1977.

3404 - SCHMIDT, L., KEC, V.: Závlaha cukrovky ve světě a v ČSR. [Irrigation of sugar beet in the world and in ČSR.] - Listy cukrovarnické (Praha) 93: 41-45, 1977. [In Czech, ab: R, E, G.]

*3405 - SCHMIDT, W.: Experimental ecology. - In: ELLENBERG, H., ESSER, K., MERXMÜLLER, H., SCHNEPF, E., ZIEGLER, H. (ed.): Progress in Botany. Morphology, Physiology, Genetics, Taxonomy, Geobotany 38. Pp. 352-366. Springer-Verlag, Berlin - Heidelberg - New York 1976.

3406 - SCHNABL, H., ZIEGLER, H.: The mechanism of stomatal movement in *Allium cepa* L. - Planta 136: 37-43, 1977.

3407 - SCHOBERT, B.: Is there an osmotic regulatory mechanism in algae and higher plants? - J. theor. Biol. 68: 17-26, 1977.

3408 - SCHOBERT, B.: The influence of water stress on the metabolism of diatoms. II. Proline accumulation under different conditions of stress and light. - Z. Pflanzenphysiol. 85: 451-461, 1977.

3409 - SCHOBERT, B.: The influence of water stress on the metabolism of diatoms. III. The effect of different nitrogen sources on proline accumulation. - Z. Pflanzenphysiol. 85: 463-470, 1977.

3410 - SCHOCH, P.G., LECHARNY, A., JACQUES, R., ZINSOU, C.: Phytochrome et indice stomatique des feuilles du *Vigna sinensis* L. - C.R. Acad. Sci. Paris, Sér. D 285: 877-879, 1977.

3411 - SCHOCH, P.G., LECHARNY, A., ZINSOU, C.: Influence de l'éclairement et de la température sur l'indice stomatique des feuilles du *Vigna sinensis* L. - C.R. Acad. Sci. Paris, Sér. D 285: 673-675, 1977.

3412 - SCHRÖDER, D., SCHEGIEWAL, A.D.: Wasserhaushalt, Bodengefüge, Ertrag, Wurzelentwicklung und Nährstoffgehalte einer Pararendzina und einer Braunerde. - Z. Acker- Pflanzenbau 145: 51-65, 1977.

3413 - SCHWARZBACH, E., KENDLBACHER, R.: Die Untersuchung früher Entwicklungsstadien des Mehltaues auf Blattoberflächen mit dem Jodabdruckverfahren. - Z. Pflanzenzücht. 78: 83-87, 1977.

3414 - SCIFRES, C.J., DURHAM, G.P., MUTZ, J.L.: Range forage production and consumption following aerial spraying of mixed brush. - Weed Sci. 25: 48-54, 1977.

3415 - SCOTT, B.I.H., GULLINE, H.F., ROBINSON, G.R.: Circadian electrochemical changes in the pulvinules of *Trifolium repens* L. - Aust. J. Plant Physiol. 4: 193-206, 1977.

3416 - SCOTT, D., HANSON, M.A.: Effect of low temperature during initial germination of some New Zealand pasture species. - N. Zeal. J. exp. Agr. 5: 41-45, 1977.

3417 - SEATON, K.A., LANDSBERG, J.J., SEDGLEY, R.H.: Transpiration and leaf water potentials of wheat in relation to changing soil water potential. - Aust. J. agr. Res. 28: 355-367, 1977.

3418 - SEPASKHAH, A.R.: Effects of soil salinity levels and plant water stress at various soybean growth stages. - Can. J. Plant Sci. 57: 925-927, 1977

3419 - SEPASKHAH, A.R.: Estimating leaf water potential in safflower. - Agron. J. 69: 894-896, 1977.

*3420 - SEPASKHAH, A.R., ARDEKANI, E.R.: Effects of irrigation regimes on yield, yield components and grain quality of Torsh barley. - Iran J. agr. Res. 4: 89-98, 1976.

3421 - SHAFFER, M.J.: Detailed return flow salinity and nutrient simulation model. -
In: DREGNE, H.E. (ed.): Managing Saline Water for Irrigarion. Pp. 127-141.
Texas Technical University, Lubbock 1977.

3422 - SHAH, G.L., DANAIAH, V., ARUNAKUMARI, P.: Observations on the effect of
Morphactin EMD 7301 and EMD 7311 on the number, size, morphology nad ontogeny
of cotyledonary stomata of *Abelmoschus esculentus* Moench. - Biol. Plant. 19:
401-404, 1977.

3423 - SHARMA, D.C., PUNTAMKAR, S.S., JAIN, S.V., SETH, S.P.: Effect of different
frequencies of irrigation with saline water on the yield of wheat and salt
accumulation in saline sodic soil. - Ind. J. agr. Sci. 47: 485-488, 1977.

*3424 - SHARMA, G.K.: Altitudinal variation in leaf epidermal patterns of *Cannabis sa-
tiva*. - Bull. Torrey bot. Club 102: 199-200, 1975.

3425 - SHARMA, M.P., VANDEN BORN, W.H., McBEATH, D.K.: Effects of post-emergence wild
oat herbicides on the transpiration of wild oats. - Can. J. Plant Sci. 57: 127
-132, 1977.

3426 - SHARMA, R.B., GHILDYAL, B.P.: Soil water-root relations in wheat: Water extrac-
tions rate of wheat roots that developed under dry and moist conditions. -
Agron. J. 69: 231-233, 1977.

*3427 - SHARMA, S.R., RAGHU, J.S., CHOUBEY, S.D.: Response of soybean varieties to ir-
rigation given at different growth stages under normal and late sown conditions. - Ind. J. Agron. 21: 412-414, 1976.

3428 - SHATILOV, I.S.: Pochvennye faktory fotosinteticheskoĭ deyatel'nosti i produktivnosti i printsipy polucheniya planiruemykh urozhaev. [Soil factors of photosynthetic activity and productivity and principles of reaching the planned
yields.] - Itogi Nauki Tekh., Ser. Fiziol. Rast. 3: 126-134, 1977. [In R.]

3429 - SHAW, R.H.: Use of moisture-stress index for examining climate trends and corn
yields in Iowa. - Iowa State J. Res. 51: 249-254, 1977.

3430 - SHAYBANY, B., MARTIN, G.C.: Abscisic acid identification and its quantitation
in leaves of *Juglans* seedlings during waterlogging. - J. Amer. Soc. hort. Sci.
102: 300-302, 1977.

3431 - SHAYKEWICH, C.F., STROOSNIJDER, L.: The concept of matrix flux potential applied to simulation of evaporation from soil. - Neth. J. agr. Sci. 25: 63-82,
1977.

3432 - SHCHERBAKOV, B.I.: Rostovye ritmy vegetativnykh organov pshenitsi i vliyanie
na nikh zasukhi. [Growth rhythms of the wheat vegetative organs and the effect
of drought on them.] - Fiziol. Rast. 24: 113-117, 1977. [In R, ab: E.]

3433 - SHEEHY, J.E., WINDRAM, A., PEACOCK, J.M.: Microclimate, canopy structure and
photosynthesis in canopies of three contrasting temperate forage grasses. II.
Microclimate and canopy structure. - Ann. Bot. 41: 579-592, 1977.

*3434 - SHEIKH, K.H., KHANUM, S.: Some studies of the quality of irrigation water and
the germination and growth of wheat at different concentrations of boron. -
Plant Soil 45: 565-576, 1976.

3435 - SHEIKHOLESLAM, S.N., CURRIER, H.B.: Phloem pressure differences and $^{14}$C-assimi-
late translocation in *Ecballium elaterium*. - Plant Physiol. 59: 376-380, 1977.

3436 - SHEIKHOLESLAM, S.N., CURRIER, H.B.: Effect of water stress on turgor differences and $^{14}$C-assimilate movement in phloem of *Ecballium elaterium*. - Plant
Physiol. 59: 381-383, 1977.

3437 - SHERIFF, D.W.: The effect of humidity on water uptake by, and viscous flow resistance of, excised leaves of a numer of species: Physiological and anatomical
observations. - J. exp. Bot. 28: 1399-1407, 1977.

3438 - SHERIFF, D.W.: Evaporation sites and distillation in leaves. - Ann. Bot. 41:
1081-1082, 1977.

3439 - SHERIFF, D.W.: Where is humidity sensed when stomata respond to it directly? -
Ann. Bot. 41: 1083-1084, 1977.

3440 - SHERIFF, D.W., KAYE, P.E.: Changes in hydraulic resistance prior to, and following illumination at the beginning of the photoperiod in *Atriplex hastata* L. - Ann. Bot. 41: 465-467, 1977.

3441 - SHERIFF, D.W., KAYE, P.E.: Responses of diffusive conductance to humidity in a drought avoiding and a drought resistant (in terms of stomatal response) legume. - Ann. Bot. 41: 653-655, 1977.

3442 - SHERIFF, D.W., KAYE, P.E.: The response of diffusive conductance in wilted and unwilted *Atriplex hastata* L. leaves to humidity. - Z. Pflanzenphysiol. 83: 463-466, 1977.

*3443 - SHINN, J.H., CLEGG, B.R., STUART, M.L., THOMPSON, S.E.: Exposures of field-grown lettuce to geothermal air pollution - photosynthetic and stomatal responses. - J. environm. Sci. Health, A11: 603-612, 1976.

*3444 - SHINOHARA, T., KITANO, H., FUKUDA, M.: [Effects of inhibitors of photosynthesis on the ozone injury to tobacco plants.] - Bull. Okayama Tobacco exp. Sta. 36: 83-86, 1976. [In Jap, ab: E.]

3445 - SHOKES, F.M., LYDA, S.D., JORDAN, W.R. Effect of water potential on the growth and survival of *Macrophomina phaseolina*. - Phytopathology 67: 239-241, 1977.

3446 - SHOMER-ILAN, A., GALUN, M., WAISEL, Y.: Seasonal variations in carbon isotope ratios in lichens. - Isr. J. Bot. 26: 46, 1977.

3447 - SHONE, M.G.T., WOOD, A.V.: Longitudinal movement and loss of nutrients, pesticides, and water in barley roots. - J. exp. Bot. 28: 872-885, 1977.

3448 - SHUKLA, A.K., BAIJAL, B.D.: Effect of salinity on IAA oxidase activity. - Ind. J. Plant Physiol. 20: 157-160, 1977.

3449 - SHUTTLEWORTH, W.J.: Comments on "resistance of a partially wet canopy: Whose equation fails?" - Boundary-Layer Meteorol. 12: 385-386, 1977.

3450 - SHUTTLEWORTH, W.J.: The exchange of wind-driven fog and mist between vegetation and the atmosphere. - Boundary-layer Meteorol. 12: 463-489, 1977.

3451 - SILVIUS, J.E., JOHNSON, R.R., PETERS, D.B.: Effect of water stress on carbon assimilation and distribution in soybean plants at different stages of development. - Crop Sci. 17: 713-716, 1977.

*3452 - ŠIMON, J.: Vliv stupňovaných dávek dusíku a závlahy na sklizeň raných kedlubnů ve středním Polabí. [Influence of gradated nitrogen doses and irrigation on early kohlrabi harvest in Central Elbe Basin.] - Zahradnictví (Praha) 2 (1): 37-45, 1975. [In Czech, ab: E, R, G.]

3453 - ŠIMON, J.: Tvorba biomasy, struktura a výše výnosu bobu obecného (*Vicia faba* L.) na lehkých pudách v závlaze. [Biomass production, structure and level of yield of horse bean (*Vicia faba* L.) on light-textured irrigated soils.] - Rost. Výroba (Praha) 23: 1169-1176, 1977. [In Czech, ab: E, G, R.]

3454 - ŠIMON, J.: Tvorba a produkce biomasy cukrovky v závlaze na lehké půdě. [Formation and production of phytomass in sugarbeet under irrigation on light soil] - In: Produkce Biomasy a Tvorba Výnosů Polních Plodin. Vol. 2. Pp. 69-77. ČVTSZ, Praha 1977. [In Czech, ab: R, E, G.]

3455 - SINGH, A.: Effect of phasic drought on the yield, water use and moisture extraction pattern of hybrid grain sorghum in Marwar tract of Rajasthan. - Ann. arid Zone 16: 231-239, 1977.

3456 - SINGH, A.: Effect of irrigation and fertilizer on the yield and quality of seed cotton (*Gossypium hirsutum*) in Western Rajasthan. - Ann. arid Zone 16: 67-72, 1977.

3457 - SINGH, A.P.: The subcellular organization of stomatal initials in sugarcane leaves: The guard and subsidiary mother cells. - Can. J. Bot. 55: 2801-2809, 1977.

3458 - SINGH, C., JACOBSON, L.: Polar movement of ions in barley roots. - Physiol. Plant. 39: 73-78, 1977.

3459 - SINGH, M., SINGH, K., SINGH, T.N.: Effect of time of first irrigation on yield
       and uptake of nitrogen in dwarf wheat varieties. - Ind. J. Agron. 22: 19-27,
       1977.

3460 - SINGH, R., CHADHA, R.K., VERMA, H.N., SINGH, Y.: Response of dryland wheat to
       phosphorus fertilizer as influenced by profile water storage and rainfall. -
       J. agr. Sci. 88: 591-595, 1977.

*3461 - SINGH, R., GILL, A.S., VERMA, H.N.: Water use and yield of dryland wheat as
       affected by N and P fertilization in loamy sand. - Ind. J. Agron. 21: 254-257,
       1976.

3462 - SINGH, R.A., SINGH, O.P., SHARMA, H.C., SINGH, M.: Effect of levels of nitrogen
       and phosphorus on yield, oil content and moisture-use pattern of rainfed win-
       ter sunflower. - Ind. J. agr. Sci. 47: 96-99, 1977.

3463 - SINGH, R.P., RAMAKRISHNA, Y.S.: Yield and moisture utilization patterns of
       dryland crops grown on conserved soil moisture. - Ann. arid Zone 16: 257-262,
       1977.

3464 - SIONIT, N.: Water status and yield of sunflowers (*Helianthus annuus*) subjected
       to water stress during four stages of development. - J. agr. Sci. 89: 663-666,
       1977.

3465 - SIONIT, N., KRAMER, P.J.: Effect of water stress during different stages of
       growth of soybean. - Agron. J. 69: 274-278, 1977.

3466 - SIPOŞ, G., PĂLTINEANU, R., AVRIGEANU, G., IARCA, P., PĂTRĂSCOIU, G., CIOTEA, V.:
       Regimul de irigare şi consumul de apă la bumbac în condiţiile Cîmpiei Române.
       [Irrigation regime and water consumption at cotton cultivated in the Romanian
       Plain.] - An. Inst. Cercetări Pentru Cereale Plante Tehnice - Fundulea 42:
       315-321, 1977. [In Roum, ab: R, E.]

3467 - SIPOS, S.: Stroenie posevnogo lozha i razvitie rasteniĭ v nachal'noĭ faze.
       [The structure of the substrate for seed and seedling development.] - In:
       Produkce Biomasy a Tvorba Výnosů Polních Plodin. Vol. 2. Pp. 174-186. ČVTSZ,
       Praha 1977. [In R, ab: Czech, G, E.]

3468 - SIROTENKO, O.D.: Ein nichtstationäres Modell für den Wasser- und Wärmehaushalt
       und für die Ertragsleistung eines Pflanzenbestandes. - In: UNGER, K. (ed.):
       Biophysikalische Analyse Pflanzlicher Systeme. Pp. 38-50. VEB Gustav Fischer
       Verlag, Jena 1977.

3469 - SIROTENKO, O.D., GORBACHEV, V.A.: Dinamicheskie modeli produktivnosti agrotse-
       nozov i problemy modelirovaniya protsessov energo- i massoobmena v sisteme
       "pochva - rastenie - atmosfera". [Dynamic models of productivity of agrocoe-
       noses and  problems of modelling processes of energy and mass exchange in the
       system "soil - plant - atmosphere".] - Itogi Nauki Tech., Ser. Fiziol. Rast.
       3: 90-107, 1977. [In R.]

3470 - SIVTSEV, M.V., DONDO, V.V.: Korrelyatsiya dinamiki soderzhaniya khlorofilla i
       aktivnosti khlorofillazy v list'yakh rasteniĭ. [Correlation of chlorophyll dy-
       namics and chlorophyllase activity in plant leaves.] - Izv. Akad. Nauk SSSR,
       Ser. biol. 77: 186-193, 1977. [In R, ab: E.]

3471 - SKOGERBOE, G.V., WALKER, W.R., EVANS, R.G., SMITH, S.W.: Evaluating improved
       irrigation water management technologies for reducing salt pickup from Grand
       Valley. - In: DREGNE, H.E. (ed.): Managing Saline Water for Irrigation. Pp.
       506-536. Texas Technical University, Lubbock 1977.

3472 - SLABBERS, P.J.: Surface roughness of crops and potential evapotranspiration. -
       J. Hydrol. 34: 181-191, 1977.

3473 - SLACK, D.C., HAAN, C.T., WELLS, L.G.: Modeling soil water movement into plant
       roots. - Trans. ASAE 20: 919-927, 933, 1977.

3474 - SLAVÍK, B.: An analogue model for the estimation of the transmesophyllar dif-
       fusive resistance for $CO_2$ in a photosynthesizing amphistomatous anisolateral
       leaf. - In: UNGER, K. (ed.): Biophysikalische Analyse Pflanzlicher Systeme.
       Pp. 146-150. VEB Gustav Fischer Verlag, Jena 1977.

3475 - **SLAVÍK, B.**: Některé zajímavé a důležité problémy vodního provozu a fyziologie fotosyntézy. [Some interesting and significant problems in water relations and physiology of photosynthesis.] - In: Dni Rastlinnej Fyziológie. Pp. 102-113, 294. Slovenská Botanická Spoločnost SAV, Bratislava 1977. [In Czech, ab: R, E.]

3476 - **SLAVÍK, L.**: Posouzení vlivu řízeného vodního režimu na tvorbu výnosu ozimé pšenice. [Appreciation of the coordinated moisture supply influence upon the farming of winter wheat yield.] - In: Produkce Biomasy a Tvorba Výnosu Polních Plodin. Vol. 2. Pp. 52-64. CVTSZ, Praha 1977. [In Czech, ab: R, E, G.]

3477 - **SLUKHAĬ, S.I., TKACHUK, E.S., KIRICHENKO, V.P., KOLOMIETS, N.G., GRINCHUK, M. A.**: Vliyanie azotnogo pitaniya na vodnyĭ rezhim ozimoĭ pshenitsy. [Effect of nitrogen nutrition on water regime of winter wheat.] - Fiziol. Biokhim. kul't. Rast. 9: 122-128, 1977. [In R, ab: E.]

3478 - **SMILES, D.E.**: Air-water-heat relationship. - In: RUSSELL, J.S., GREACEN, E.L. (ed.): Soil Factors in Crop Production in a Semi-Arid Environment. Pp. 54-77. University of Queensland Press, St. Lucia 1977.

3479 - **SMITH, W.K., NOBEL, P.S.**: Temperature and water relations for sun and shade leaves of desert broadleaf, *Hyptis emoryi*. - J. exp. Bot. 28: 169-183, 1977.

3480 - **SMITH, W.K., NOBEL, P.S.**: Influences of seasonal changes in leaf morphology on water-use efficiency for three desert broadleaf shrubs. - Ecology 58: 1033-1043, 1977.

3481 - **SMOLYAK, L.P., RUSALENKA, A.I.**: Uplyŭ padtaplennya i zataplennya na radyyal'ny pryrost sasny zvychaĭnaĭ. [Effect of flooding on radial growth of pine.] - Vestsi Akad. Navuk Belarus. SSR, Ser. biyal. Navuk 1977 (5): 20-25, 1977.[In Belorus.]

*3482 - **SMUKALSKI, M., MORITZ, L., ROGMANN, H.**: Wechselbeziehungen von Klarwasserberegnung, Unterbodenerschliessung und hohem Mineraldüngergaben auf Sandboden. - Arch. Acker- Pflanzenbau Bodenk. 20: 567-579, 1976.

3483 - **SMUKALSKI, M., ROGASIK, J.**: Einfluss von Beregnung und Intensivdüngung auf Ertragsverhalten und Nährstoffaufnahme von Zuckerrüben in Fruchtfolge und Monokultur. - Arch. Acker- Pflanzenbau Bodenk. 21: 659-673, 1977.

3484 - **SNOECK, J.**: Essai d'irrigation du caféier Robusta. - Café Cacao Thé 21: 111-125, 1977.

3485 - **SOFIELD, I., WARDLAW, I.F., EVANS, L.T., ZEE, S.Y.**: Nitrogen, phosphorus and water content during grain development and maturation in wheat. - Aust. J. Plant Physiol. 4: 799-810, 1977.

3486 - **SOJKA, R.E., SCOTT, H.D., FERGUSON, J.A., RUTLEDGE, E.M.**: Relation of plant water status to soybean growth. - Soil Sci. 123: 182-187, 1977.

3487 - **SOJKA, R.E., STOLZY, L.H., KAUFMANN, M.R.**: Wheat growth related to rhizosphere temperature and oxygen levels. - Agron. J. 67. 501-596, 1977.

3488 - **SOLÁROVÁ, J., VÁCLAVÍK, J., POSPÍŠILOVÁ, J.**: Leaf conductance and gas exchange through adaxial and abaxial surfaces in water stressed primary bean leaves. - Biol. Plant. 19: 59-64, 1977.

3489 - **SOLÁROVÁ, J., VÁCLAVÍK, J., TICHÁ, I., ČATSKÝ, J.**: Vliv hustoty ozáření během kultivace na výměnu plynů u primarních listů fazolu. [The effect of growth irradiance on gas exchange of primary bean leaves.] - In: Dni Rastlinnej Fyziológie. Pp. 130-133, 299. Slovenská Botanická Spoločnost SAV, Bratislava 1977. [In Czech, ab: R, E.]

3490 - **SOLÁROVÁ, J., ZIMA, J., VÁCLAVÍK, J., ŠESTÁK, Z.**: Effect of irradiance during growth of French bean on some characteristics of photosynthesis and water relations. - In: Pigment-Protein Complexes in Photosynthesis. Biological Research Center, Szeged 1977.

3491 - **ŠPÁNIK, F., KRÁL, M.**: Možnosť sledovania vodného stavu rastlín ozimnej pšenice konduktometrickou metodou. [Possibility of determination of water status of winter wheat plants by conductometric method.] - In: REPKA, J. (ed.): Zborník Referátov zo Seminára Fyziologicko-Genetické a Chemické Faktory Produktivity Rastlín. Pp. 171-178. Vysoká Škola Poľnohospodárská, Nitra 1977. [In Slov.]

*3492 - SPECTY, R.: Étude des effets cumulés et des arrière-effets de plusieurs fac-
        teurs agronomiques appliqués dans le cadre de rotations possibles dans la hardt
        irriguée. I. Résultats partiels sur cultures de maïs grain. - Ann. agron. 26:
        169-197, 1975.

 3493 - SPIERTZ, J.H.J.: The influence of temperature and light intensity on grain
        growth in relation to the carbohydrate and nitrogen economy of the wheat plant.
        - Neth. J. agr. Sci. 25: 182-197, 1977.

 3494 - SPRENT, J.I., BRADFORD, A.M.: Nitrogen fixation in field beans (*Vicia faba*) as
        affected by population density, shading and its relationship with soil mois-
        ture. - J. agr. Sci. 88: 303-310, 1977.

 3495 - SPRENT, J.I., BRADFORD, A.M., NORTON, C.: Seasonal growth patterns in field
        beans (*Vicia faba*) as affected by population density, shading and its rela-
        tionship with soil moisture. - J. agr. Sci. 88: 293-301, 1977.

*3496 - SPURWAY, R.A., HEDGES, D.A., WHEELER, J.L.: The quality and quantity of forage
        oats sown at intervals during autumn: Effects of nitrogen and supplementary
        irrigation. - Aust. J. exp. Agr. anim. Husb. 16: 555-563, 1976.

 3497 - SQUIRE, G.R.: Seasonal changes in photosynthesis of tea (*Camellia sinensis* L.)
        - J. appl. Ecol. 14: 303-316, 1977.

 3498 - SRINIVAS, T., BHASHYAM, M.K., MAHADEVAPPA, M., DESIKACHAR, H.S.R.: Varietal
        differences in crack formation due to weathering and wetting stress in rice. -
        Ind. J. agr. Sci. 47: 27-31, 1977.

*3499 - SRIVASTAVA, H.S.: Inhibition of maize seedling growth by chloramphenicol. -
        Ind. J. Plant Physiol. 19: 53-59, 1976.

 3500 - STADELMANN, E.J.: Passive transport parameters of plant cell membranes. - In:
        MARRÉ, E., CIFERRI, O. (ed.): Regulation of Cell Membrane Activities in Plants.
        Pp. 3-18. Elsevier/North-Holland Biomedical Press, Amsterdam - Oxford - New
        York 1977.

 3501 - STANHILL, G.: Quantifying weather-crop relations. - In: LANDSBERG, J.J.,
        CUTTING, C.V. (ed.): Environmental Effects on Crop Physiology. Pp. 23-37. Aca-
        demic Press, London - New York - San Francisco 1977.

 3502 - STEGMAN, E.C., BAUER, A.: Sugar beet response to water stress in sandy soils.
        - Trans. ASAE 20: 469-473, 477, 1977.

 3503 - STELZER, R., LÄUCHLI, A.: Salz- und Überflutungstoleranz von *Puccinellia pei-
        sonis*. II. Structurelle Differenzierung der Wurzel in Beziehung zur Funktion. -
        - Z. Pflanzenphysiol. 84: 95-108, 1977.

 3504 - ŠTĚPÁNEK, V.: Vodní provoz u adultního stromu dubu letního (*Quercus robur* L.)
        [Water relations in adult tree of oak (*Quercus robur* L.).]- In: HUZULÁK, J. MA-
        SAROVIČOVÁ, E. (ed.) Fotosyntéza a Vodný Režim Drevín. Pp. 259-263. Modra-
        Piesky 1977. [In Czech, ab: E, R.]

 3505 - STERNE, R.E., KAUFMANN, M.R., ZENTMYER, G.A.:  Environmental effects on tran-
        spiration and leaf water potential in avocado. - Physiol. Plant. 41: 1-6, 1977.

*3506 - STERNE, R.E., ZENTMYER, G.A., BINGHAM, F.T.: The effect of osmotic potential
        and specific ions on growth of *Phytophthora cinnamomi*. - Phytopathology 66:
        1398-1402, 1976.

 3507 - STERNE, R.E., ZENTMYER, G.A., KAUFMANN, M.R.: The effect of matric and osmotic
        potential of soil on *Phytophthora* root disease of *Persea indica*. - Phytopa-
        thology 67: 1491-1494, 1977.

 3508 - STERNE, R.E., ZENTMYER, G.A., KAUFMANN, M.R.: The influence of matric potential,
        soil texture, and soil amendment on root disease caused by *Phytophthora cinna-
        momi*. - Phytopathology 67: 1495 - 1500, 1977.

 3509 - STEUDLE, E., ZIMMERMANN, U., LÜTTGE, U.: Effect of turgor pressure and cell
        size on the wall elasticity of plant cells. - Plant Physiol. 59: 285-289, 1977.

 3510 - STEVENS, E., STEVENS, L.: Glucose-6-phosphate dehydrogenase activity under con-
        ditions of water limitation: A possible model system for enzyme reactions in un-
        imbibed resting seeds and its relevance to seed viability. - J. exp. Bot. 28:
        292-303, 1977.

3511 - STEVENS, R.A., MARTIN, E.S.: The morphogenesis of substomatal structures in *Polypodium vulgare*. - Can. J. Bot. 55: 2873-2878, 1977.

3512 - STEVENS, R.A., MARTIN, E.S.: New structure associated with stomatal complex of the fern *Polypodium vulgare*. - Nature 265: 331-334, 1977.

3513 - STEVENS, R.A., MARTIN, E.S.: Ion-adsorbent substomatal structures in *Tradescantia pallidus*. - Nature 268: 364-365, 1977.

3514 - STEWART, C.R., BOGGESS, S.F.: The effect of wilting on the conversion of arginine, ornithine, and glutamate to proline in bean leaves. - Plant Sci. Lett. 8: 147-153, 1977.

3515 - STEWART, C.R., BOGGESS, S.F., ASPINALL, D., PALEG, L.G.: Inhibition of proline oxidation by water stress. - Plant Physiol. 59: 930-932, 1977.

3516 - STEWART, J.I.: Conservation irrigation of field crops: a drought-year strategy. - California Agr. 31: 6-9, 1977.

3517 - STEWART, J.I., HAGAN, R.M., PRUITT, W.O.: Salinity effects on corn yield, evapotranspiration, leaching fraction, and irrigation efficiency. - In: DREGNE, H.E. (ed.): Managing Saline Water for Irrigation. Pp. 316-332. Texas Technical University, Lubbock 1977.

3518 - STIBBE, E., HADAS, A.: Response of dryland cotton plant growth, soil-water uptake, and lint yield to two extreme types of tillage. - Agron. J. 69: 447-451, 1977.

3519 - STIGTER, C.J.: Water vapour pressure within a maize crop. - Arch. Meteorol. Geophys. Bioklimatol., Ser. B, 24: 349-359, 1977.

3520 - STIGTER, C.J., GOUDRIAAN, J., BOTTEMANNE, F.A., BIRNIE, J., LENGKEEK, J.G., SIBMA, L.: Experimental evaluation of a crop climate simulation model for Indian corn (*Zea mays* L.). - Agr. Meteorol. 18: 163-186, 1977.

*3521 - STIGTER, C.J., LENGKEEK, J.G., KOOIJMAN, J.: A simple worst case analysis for estimation of correct scanning rate in a micrometeorological experiment. - Neth. J. agr. Sci. 24: 3-16, 1976.

3522 - STOKER, R.: Irrigation of garden peas on a good cropping soil. - N. Zeal. J. exp. Agr. 5: 233-236, 1977.

3523 - STOM, D.I., ROGOZINA, N.A.: K voprosu o vliyanii na kul'turu svekly fenolov pochvy i produktov ikh okisleniya. [Influence of soil phenols and products of oxydation on sugar beet culture.] - Sel'skokhoz. Biol. 12: 462-464, 1977. [In R.]

3524 - STOUT, D.G., COTTS, R.M., STEPONKUS, P.L.: The diffusional water permeability of *Elodea* leaf cells as measured by nuclear magnetic resonance. - Can. J. Bot. 55: 1623-1631, 1977.

3525 - STOUT, D.G., STEPONKUS, P.L., COTTS, R.M.: Quantitative study of importance of water permeability in plant cold hardiness. - Plant Physiol. 60: 374-378, 1977.

3526 - STREBEYKO, P.: Susza 1976 r. w Europie zachodniej. [Drought in Western Europe in 1976.] - Postepy Nauk rolnicz. 24(6): 29-46, 1977. [In Pol.]

3527 - STUART, D.A., JONES, R.L.: Roles of extensibility and turgor in gibberellin- and dark-stimulated growth. - Plant Physiol. 59: 61-68, 1977.

3528 - STUMM, G.: Wirtschaftlichkeit der Beregnung im deutschen Weinbau unter dem Aspekt der Mehrzweckberegnung. - Ber. Landwirt. 55: 64-78, 1977.

3529 - STURGES, D.L.: Soil water withdrawal and root characteristics of big sagebrush. - Amer. Midl. Nat. 98: 257-274, 1977.

3530 - SUAREZ, J.J., HERNANDEZ, A.: Effects of soil water deficit on the biological indicators of guinea grass (*Panicum maximum* Jacq.) and glycine (*Glycine wightii*) pastures. - Cuban J. agr. Sci. 11: 217-225, 1977.

3531 - SUD'INA, E.G., DOVBYSH, E.F., GOLOD, M.G., DONTSOVA, I.G., BAĬDULOVA-BABKO, T. Yu.: Khlorofillaza v izolirovannykh list'yakh tabaka. [Chlorophyllase in isolated tobacco leaves.] - Fiziol. Biokhim. kul't. Rast. 9: 418-423, 1977. [In R, ab: E.]

*3532 - SUDO, K., ANDO, T.: [Influence of atmospheric humidity and soil moisture con-
         tents on the plant water condition  as well as on the growth of sweet pepper
         and tomato plants.] - Bull. vegetable ornament. Crops Res. Sta. Ser. A 1975
         (2): 49-63, 1975. [In Jap, ab: E.]

*3533 - SUGIMOTO, K.: [Studies on evapo-transpiration of indica rice plants with ref-
         erence to crop science.] - Jap. J. trop. Agr. 20: 33-34, 1976. [In Jap.]

 3534 - SUGIMOTO, K., SASIPRAPA, V., BANGLIANG, S.: [Effects of irrigation and ferti-
         lization on some upland crops in paddy field of Thailand.] - Jap. J. Crop Sci.
         46: 119-124, 1977. [In Jap, ab: E.]

 3535 - SUMAYAO, C.R., KANEMASU, E.T., HODGES, T.: Soil moisture effects on transpira-
         tion and net carbon dioxide exchange of sorghum. - Agr. Meteorol. 18: 401-408,
         1977.

 3536 - SUTCLIFFE, J.: Plants and Temperature. (Studies in Biology 86.) Edward Arnold
         (Publ.) Ltd., London 1977.

 3537 - SUTCLIFFE, J.F., PATE, J.S. (ed.): The Physiology of the Garden Pea. (Experi-
         mental Botany: An International Series of Monographs Vol. 12). Academic Press,
         London - New York - San Francisco 1977.

 3538 - ŠÚTOR, J.: Súčasné aspekty výzkumu tvorby vlhkostného režimu pôdy. [The pre-
         sent research aspects of the soil moisture regime formation.] - In: HUZULÁK,
         J., MASAROVIČOVÁ, E. (ed.): Fotosyntéza a Vodný Režim Drevín. Pp. 135-141.
         Modra-Piesky 1977. [In Slov, ab: E, R.]

 3539 - ŠVACHULA, V., KOHOUT, V.: Porovnání metabolismu cukrové řepy a merlíku bílého
         v diferencovaných vláhových podmínkách. [Comparison of metabolism of sugar
         beet and dungweed under differentiated moisture conditions.] - Rost. Výroba
         23: 77-88, 1977. [In Czech, ab: R, E.]

 3540 - ŠVACHULA, V., ŠVACHULOVÁ, J.: Die Erhöhung der Peroxidasenaktivität, des Chlo-
         rophyllgehaltes und der Zuckerproduktion in der Zuckerrübe in bewässerter
         Fruchtfolge. - Z. Acker- Pflanzenbau 145: 142-153, 1977.

 3541 - ŠVACHULA, V., TORNIKIDU, J., ZAHRADNÍČEK, J.: Vliv vědecky řízené závlahy cuk-
         rovky na její technologickou jakost a posklizňový metabolismus. [Influence of
         scientific control of irrigation of sugar beet on its technological quality
         and on metabolism after the harvesting.] - Listy cukrovarnické (Praha) 93: 35-
         41, 1977. [In Czech, ab: R, E, G.]

 3542 - SVENSSON, B.: Changes in seed tubers after planting. - Potato Res. 20: 215-218,
         1977.

 3543 - ŠVIHRA, J., HOJČUŠ, R.: Saturácia pletív vodou a produkčný proces obilnín.
         [The hydration of tissues in relation to the growth and production processes
         in cereals.] - In: Dni Rastlinnej Fyziológie. Pp. 158-163, 306. Slovenská Bo-
         tanická Spoločnost SAV, Bratislava 1977. [In Slov, ab: R, E.]

 3544 - ŠVIHRA, J., ZIMA, M., HOJČUŠ, R.: Vplyv vodného stresu v období po 7. etape
         organogenézy na produktívnosť ozimnej pšenice. [Effect of water stress after
         7th stage of organogenesis on productivity of winter wheat.] - Rost. Výroba
         23: 675-681, 1977. [In Slov, ab: R, E, G.]

 3545 - SVOBODOVÁ, J.: Výnosové a kvalitativní ukazatele sklizní na odvodněné černoze-
         mi - smonici. [The yield and quality indices of harvests on drained chernozem
         - smonitza soil.] - Rost. Výroba 23: 379-389. 1977. [In Czech, ab: R, E, G.]

 3546 - SYBER, A.Yu., MOLDAU, Kh.A.: Apparatura dlya opredeleniya soprotivleniya
         ust'its i vodosoderzhaniya lista s razdel'nym konditsionirovaniem rasteniya i
         otdel'nogo lista. [An apparatus with separate conditioning of the plant and
         the leaf for the determination of stomatal resistance and leaf water content.]
         - Fiziol. Rast. 24: 1301-1307, 1977. [In R, ab: E.]

 3547 - SZANIAWSKI, R., ŻELAWSKI, W., WIERZBICKI, B.: Wymiana gazowa i gospodarka wod-
         na. [Gas exchange and water relations.] - In: BIALOBOK, S. (ed.) Swierk pospo-
         lity - Picea abies (L.) Karst. (Nasze drzewa lesne - monografie popularnonauko-
         we.) Pp. 131-152. Institut Dendrologii PAN Kórnik, Panstwowe Wydawnictwo Nauko-
         we, Warszawa 1977. [In Pol, ab: E.]

3548 - SZANIAWSKI, R.K.: Diurnal and seasonal patterns of gas exchange in Scots pine
(*Pinus silvestris* L.) seedlings grown under laboratory conditions. - In: HUZU-
LÁK, J., MASAROVIČOVÁ, E. (ed.): Fotosyntéza a Vodný Režim Drevín. Pp. 251-258.
Modra-Piesky 1977.

3549 - SZAREK, S.R., WOODHOUSE, R.M.: Ecophysiological studies of sonoran desert
plants. II. Seasonal photosynthesis patterns and primary production of *Ambrosia deltoidea* and *Olneya tesota*. - Oecologia 28: 365-375, 1977.

*3550 - TAJIMA, K., AKITA, S., SHIMIZU, N.: [Effect of high soil temperature on growth
and water balance of tall fescue and perennial ryegrass.]- J. Jap. Soc. Grassland Sci. 22: 256-260, 1976. [In Jap, ab: E.]

3551 - TAKANO, T.: Leaf water deficits in Satsuma mandarin trees (*Citrus unshiu* Marc.)
at different seasons of the year. - Sci. Rep. Fac. Agr., Meijo Univ. 13: 18-27,
1977.

3552 - TAKEDA, T., AGATA, W., HAKOYAMA, S., TANAKA, H.: [Studies on weed vegetation
in non-cultivated paddy fields. II. The relation between the ecological distribution of gramineous $C_3$- and $C_4$-weeds and the soil moisture condition in
non-cultivated paddy fields.] - Jap. J. Crop Sci. 46: 558-568, 1977. [In Jap,
ab: E.]

*3553 - TALSMA, T., van der LELIJ, A.: Water balance estimates of evaporation from
ponded rice fields in semi-arid region. - Agr. Water Manage. 1: 89-97, 1976.

3554 - TAMM, C.O., COWLING, E.B.: Acidic precipitation and forest vegetation. - Water,
Air, Soil Pollut. 7: 503-511, 1977.

*3555 - TAN, C.S., BLACK, T.A.: Factors affecting the canopy resistance of a Douglas-
fir forest. - Boundary-Layer Meteorol. 10: 475-488, 1976.

3556 - TERRY, N.: Photosynthesis, growth, and role of chloride. - Plant Physiol. 60:
69-75, 1977.

3557 - TESHA, A.J., KUMAR, D.: Water movement in coffee seedlings (*Coffea arabica* L.).
Evaluation of various resistances. - Oecologia 28: 377-382, 1977.

3558 - THALOUARN, P., HELLER, R.: Action du chlorure mercurique sur l'imbibition des
graines de *Pinus halepensis* Mill. - Physiol. vég. 15: 1-14, 1977.

3559 - THOMAS, A.W., DUKE, H.R.: Capillary potential distributions in root zones
using subsurface irrigation. - Trans. ASAE 20: 62-67, 75, 1977.

3560 - THOMAS, D.A., REBELLA, C., CHARTIER, P.: An analysis of the vertically reflec-
ted radiation from a maize crop as a possible means of determining its biomass
and water content. - Agr. Meteorol. 18: 101-114, 1977.

3561 - THOMAS, J.R., GAUSMAN, H.W.: Leaf reflectance vs. leaf chlorophyll and caro-
tenoid concentrations for eight crops. - Agron. J. 69: 799-802, 1977.

3562 - THOMAS, R.J.: Water relations of elongating seta cells in sporophytes of the
liverwort *Lophocolea heterophylla*. - Bryologist 00: 345 347, 1977.

3563 - THOMPSON, D.R., HINCKLEY, T.M.: A simulation of water relations of white oak
based on soil moisture and atmospheric evaporative demand. - Can. J. Forest
Res. 7: 400-409, 1977.

3564 - THOMPSON, D.R., HINCKLEY, T.M.: Effect of vertical and temporal variations in
stand microclimate and soil moisture on water status of several species in an
oak-hickory forest. - Amer. Midl. Nat. 97: 373-380, 1977.

3565 - THOMPSON, J.A.: Effect of irrigation termination on yield of soybeans in
southern New South Wales. - Aust. J. exp. Agr. anim. Husb. 17: 156-160, 1977

3566 - THOR, E., CORE, H.A.: Fertilization, irrigation, and site factor relationships
with growth and wood properties of yellow-poplar. - Wood Sci. 9: 130-135, 1977.

3567 - THORPE, N., MILTHORPE, F.L.: Stomatal metabolism: $CO_2$ fixation and respiration.
- Aust. J. Plant Physiol. 4: 611-621, 1977.

3568 - TICHÁ, I., ČATSKÝ, J.: Regulace fotosyntetické aktivity listu fazolu jeho ana-
tomickou·strukturou a transportem $CO_2$ v mezofylu. [Photosynthetic activity of a
bean leaf as controlled by its anatomical structure and $CO_2$ transport in the

mesophyll.] - In: Dni Rastlinnej Fyziológie. Pp. 118-121, 296. Slovenská Botanická Spoločnost SAV, Bratislava 1977. [In Czech, ab: R, E.]

3569 - TING, I.P., HANSCOM, Z.,III.: Induction of acid metabolism in *Portulacaria afra*. - Plant Physiol. 59: 511-514, 1977.

3570 - TINGEY, D.T., STOCKWELL, C.: Semipermeable membrane system for subjecting plants to water stress. - Plant Physiol. 60: 58-60, 1977.

*3571 - TKACHUK, K.S., GULYAEV, B.I., PETRENKO, N.I.: Vliyanie azotnogo pitaniya na vodoobmen, fotosintez i produktivnost' ozimoĭ pshenitsy. [Influence of nitrogen nutrition on water exchange, photosynthesis and productivity of winter wheat.] - Dokl. Akad. Nauk Ukr. SSR, Ser. B 1975 (7): 650-653, 1975.[In R,ab:E.]

3572 - TORRES, G.A.: Spotreba vody u citrusových rastlín pestovaných pri rôznom obsahu vody a živín v pôde. [Water consumption of citrus plants grown at different soil moisture and nutrient levels.] - In: HUZULÁK, J., MASAROVIČOVÁ, E. (ed.): Fotosyntéza a Vodný Režim Drevín. Pp. 130-134. Modra-Piesky 1977. [In Slov, ab: E, R.]

3573 - TRAFFORD, B.D.: Field drainage in the humid zone. - Span 20: 81-83, 1977.

3574 - TRAN DANG HONG, MINCHIN, F.R., SUMMERFIELD, R.J.: Recovery of nodulated cowpea plants (*Vigna unguiculata* (L.) Walp.) from waterlogging during vegetative growth. - Plant Soil 48: 661-672, 1977.

3575 - TRAVIS, A.J., MANSFIELD, T.A.: Studies of malate formation in "isolated" guard cells. - New Phytol. 78: 541-546, 1977.

3576 - TROMBLE, J.M.: Water requirements for mesquite (*Prosopis juliflora*). - J. Hydrol. 34: 171-179, 1977.

*3577 - TRUKSA, J., ZÁBORSKÝ, J.: Vplyv šírky riadku a hustoty porastu pri závlahe na produkciu kukurice. [The effect of row spacing and stand density in irrigated plots on maize production.] - Rost. Výroba 22: 1041-1046, 1976. [In Czech, ab: E, R.]

*3578 - TSCHAKALOVA, E., HOFFMANN, P.: Strukturelle und funktionelle Grundlagen des photosynthetischen Gaswechsels bei *Triticum aestivum* L. - Wiss. Z. Humboldt-Univ. Berlin, math.-naturwiss. Reihe 25: 723-736, 1976.

3579 - TU, C.M., HIETKAMP, G.: Effect of moisture on acetylene reduction (symbiotic nitrogen fixation) by *Rhizobium japonicum* and soybean root nodules in silica sand. - Commun. Soil Sci. Plant Anal. 8: 81-86, 1977.

3580 - TUCKER, C.J.: Spectral estimation of grass canopy variables. - Remote Sensing Environ. 6: 11-26, 1977.

3581 - TUCKER, C.J.: Asymptotic nature of grass canopy spectral reflectance. - Appl. Optics 16: 1151-1156, 1977.

3582 - TUCKER, C.J., GARRATT, M.W.: Leaf optical system modeled as a stochastic process. - Appl. Optics 16: 635-642, 1977.

*3583 - TUCKER, C.J., MAXWELL, E.L.: Sensor design for monitoring vegetation canopies. - Photogram. Eng. remote Sensing 42: 1399-1410, 1976.

3584 - TUCKER, C.J., MILLER, L.D.: Soil spectra contributions to grass canopy spectral reflectance. - Photogram. Eng. remote Sensing 43: 721-726, 1977.

3585 - TURNER, N.C., HEICHEL, G.H.: Stomatal development and sesonal changes in diffusive resistance of primary and regrowth foliage of red oak (*Quercus rubra* L.) and red maple (*Acer rubrum* L.) - New Phytol. 78: 71-81, 1977.

*3586 - TYAGI, N.K., ACHARYYA, N., MOHANTY, P.C.: Effect of puddling implements on percolation losses and water-use efficiency in rice field. - Ind. J. agr. Sci. 45: 132-135, 1975.

3587 - TYREE, M.T., CAMERON, S.I.: A new technique for measuring oscillatory and diurnal changes in leaf thickness. - Can. J. Forest Res. 7: 540-544, 1977.

3588 - TYREE, M.T., CHEUNG, Y.N.S.: Resistance to water flow in *Fagus grandifolia* leaves. - Can. J. Bot. 55: 2591-2599, 1977.

*3589 - UCHIJIMA, Z.: Water consumption in crop production. - In: Science for Better Environment. Proceeding of the International Congress on the Human Environment. Pp. 184-193. HESC, Kyoto, 1976.

*3590 - UCHIJIMA, Z.: Microclimate of the rice crop. - In: Climate and Rice. Pp. 115-140. International Rice Research Institute, Los Baños 1976.

3591 - ÚLEHLA, J., ČERMÁK, J.: Denní úhrny transpiračního toku v kmenech dospělých stromu a potenciální evapotranspirace. [Daily totals of transpiration flow in adult trees and potential evapotranspiration.] - In: HUZULÁK, J., MASAROVIČOVÁ, E. (ed.): Fotosyntéza a Vodný Režim Drevín. Pp. 55-62. Modra-Piesky 1977. [In Czech, ab: E, R.]

*3592 - ULRICH, P.-C., WICKE, H.-J.: Zur Steuerung der Beregnung von Getreide mit Hilfe der Refraktometermethode. - Tagungsber. Akad. Landwirtschaftswiss. DDR, Berlin 135: 243-251, 1975.

3593 - UNGAR, I.A.: The relationship between soil water potential and plant water potential in two inland halophytes under field conditions. - Bot. Gaz. 138: 498-501, 1977.

3594 - UNGAR, I.A.: Salinity, temperature, and growth regulator effects on seed germination of *Salicornia europaea* L. - Aquatic Bot. 3: 329-335, 1977.

3595 - UNGER, K. (ed.): Biophysikalische Analyse Pflanzlicher Systeme. - VEB Gustav Fischer Verlag, Jena 1977.

3596 - UNGER, P.W.: Tillage effects on winter wheat production where irrigated and dryland crops are alternated. - Agron. J. 69: 944-950, 1977.

*3597 - UNSWORTH, M.H., BISCOE, P.V., BLACK, V.: Analysis of gas exchange between plants and polluted atmospheres. - In: MANSFIELD, T.A. (ed.): Effects of Air Pollutants on Plants. Pp. 5-16. Cambridge University Press, Cambridge - London - New York - Melbourne 1976.

3598 - UPHAUS, R.A., BLAKE, M.I., KOSTKA, A.G., KATZ, J.J.: Automated growth chambers for production of isotopically substituted higher plants. - Photosynthetica 11: 314-321, 1977.

3599 - VACHNADZE, G.S., BEROSHVILI, V.G.: Vliyanie urovnya gruntovykh vod na sostoyanie i rost lesnykh kul'tur. [The influence of ground water level on the growth and the state of forest cultures.] - In: Voprosy Gornogo Lesovedeniya i Lesovodstva v Gruzii. Vol. 25. Pp. 45-51. Sabchota Abzhara, Batumi 1977. [In R.]

3600 - VÁCLAVÍK, J.: Distribution pattern of gas exchange in the area of maize leaf blades during the generative phase. - Biol. Plant. 19: 457-461, 1977.

3601 - VÁCLAVÍK, J.: Vliv stáří amfistomatických listů na fotosyntetický příjem $CO_2$ a transpiraci. [The effect of amphistomatous leaf age on net photosynthetic $CO_2$ uptake and transpiration.] - In: Dni Rastlinnej Fyziológie. Pp. 126-129, 298. Slovenská Botanická Spoločnost SAV, Bratislava 1977. [In Czech, ab: R, E.]

3602 - VALANCOGNE, C., DAUDET, F.A.: Etude in situ du potentiel hydrique sous une culture de maïs à l'aide d'un système automatique de mesure à micropsychromètre à effet Peltier. - Ann. agron. 28: 137-157, 1977.

*3603 - VAN BUIJTENEN, J.P., ZIMMERMAN, R.H.: Morpho-physiological characteristics related to drought resistance in *Pinus taeda*. - In: CANNELL, M.G.R., LAST, F.T. (ed.): Tree Physiology and Yield Improvement. Pp. 349-359. Academic Press, London - New York - San Francisco 1976.

*3604 - VAN SCHAIK, J.C.: Criteria for supplementary irrigation. - In: Efficiency of Water and Fertilizer Use in Semi-Arid Regions. A Technical Document. Pp. 229-241. International Atomic Energy Agency, Vienna 1976.

*3605 - VAN SCHILFGAARDE, J.: Some observations pertinent to irrigation management. - In: Efficiency of Water and Fertilizer Use in Semi-Arid Regions. A Technical Document. Pp. 243-248. International Atomic Energy Agency, Vienna 1976.

3606 - VAN VOLKENBURGH, E., DAVIES, W.J.: Leaf anatomy and water relations of plants grown in controlled environments and in the field. - Crop Sci. 17: 353-358, 1977.

3607 - VARLEV, I.: Evaluation of nonuniformity in irrigation and yield. - J. Irrig. Drain. Div. ASCE 102: 149-164, 1976.

3608 - VARMA, S.K.: Nutrient absorption under soil moisture stress and nitrogen deficiency in wheat (*Triticum aestivum*). - Ind. J. agr. Res. 10: 174-178, 1976.

3609 - VARMA, S.K., MALIK, B.S., AGARWAL, M.C.: Effect of nitrogen application under varying soil moisture regimes on the uptake of nutrients in barley. - Ind. J. Agron. 21: 318-319, 1976.

3610 - VARMA, S.K., MALIK, B.S., NATH, J.: Effect of soil moisture regimes, phosphorus levels and their interactions on N, P and K content and uptake in peas (*Pisam sativum* L.) under semi arid conditions. - Trans. Isdt. Ucds. 2: 111-115, 1977.

3611 - VARSHNEY, K.A., BAIJAL, B.D.: Effect of salt-stress on chlorophyll contents of some grasses. - Ind. J. Plant Physiol. 20: 161-163, 1977.

3612 - VARSHNEY, K.A., BAIJAL, B.D.: Effect of salt stress on seed germination of some pasture grasses. - Comp. Physiol. Ecol. 2: 104-106, 1977.

3613 - VARSHNEY, K.A., BAIJAL, B.D.: Note on the influence of salinity on early seedling growth of some pasture grasses. - Ind. J. agr. Res. 11: 59-61, 1977.

3614 - VASILIU, M.: Consumul de apă la soia şi sursele principale de acoperire a acestuia în condiţiile Cîmpiei Brăilei. [Soybean water consumption and its main sources in the Braila Plain.] - An. Inst. Cercetări Pentru Cereale Plante Tehnice - Fundulea 42: 281-287, 1977. [In Roum, ab: R, E.]

3615 - VASILIU, M., VASILIU, M., PASCARU, E.: Influenta irigatiei si îngrăşămintelor asupra producţiei şi calităţii recoltei se soia în Cîmpia Brăilei. [The influence of irrigation and fertilizers upon the yield and quality of soybean crop in Braila Plain.]- An. Inst. Cercetări Pentru Cereale Plante Tehnice - Fundulea 42: 299-306, 1977. [In Roum, ab: R, E.]

*3616 - VAUCLIN, M., HAVERKAMP, R., VAHCAUD, G.: Transfer hydriques dans le systeme sol-plante-atmosphere. Simulation et prevision. - In: Efficiency of Water and Fertilizer Use in Semi-Arid Regions. A Technical Document. Pp. 195-222. International Atomic Energy Agency, Vienna 1976.

3617 - VEEN, B.W.: The uptake of potassium, nitrate, water, and oxygen by a maize root system in relation to its size. - J. exp. Bot. 28: 1389-1398, 1977.

3618 - VELIKANOV, G.A., GORDON, L.Kh., VOLKOV, V.Ya., BARYSHEVA, T.S.: Izuchenie pronitsaemosti kletochnykh membran dlya vody in vivo metodom YaMR. [Study of cell membranes permeability for water in vivo by the NMR method.] - Fiziol. Biokhim. kul't. Rast. 9: 197-201, 1977. [In R, ab: E.]

*3619 - VERMA, B.S., SINGH, R.R.: Effect of nitrogen, moisture regime and plant density on grain yield and quality of hybrid maize. - Ind. J. Agron. 21: 441-445, 1976.

3620 - VERMA, S.B., ROSENBERG, N.J.: Brown-Rosenberg resistance model of crop evapotranspiration modified tests in an irrigated sorghum field. - Agron. J. 69: 332-335, 1977.

3621 - VIANE, R., VAN COTTHEM, W.: Spore morphology and stomatal characters of some Kenyan *Asplenium*-species. - Ber. Deut. bot. Ges. 90: 219-239, 1977.

3622 - VICHERKOVÁ, M., KOSTŘICA, P.: Vliv draslíku na otevírání průduchů u bobu (*Vicia faba* L.) kultivovaného při dostatečné a deficitní draslíkové výživě. [Effect of potassium on stomatal opening in broad bean (*Vicia faba* L.) cultivated under sufficient and deficient potassium nutrition.] - In: Dni Rastlinnej Fyziológie. Pp. 86-88, 291. Slovenská Botanická Spoločnost SAV, Bratislava 1977. [In Czech, ab: R, E.]

*3623 - VIETS, F.G., Jr.: Efficiency of water use on semi-arid land. - In: Efficiency of Water and Fertilizer Use in Semi-Arid Regions. A Technical Document. Pp. 123-133. International Atomic Energy Agency, Vienna 1976.

3624 - VIGNES, D., CARLES, J.: Influence du vent sur l'activité photosynthétique et les échanges gazeux. - Oecol. Plant. 12: 149-158, 1977.

3625 - VIIL, J., LAISK, A., OJA, V., PÄRNIK, T.: Enhancement of photosynthesis caused by oxygen under saturating irradiance and high $CO_2$ concentrations. - Photosynthetica 11: 251-259, 1977.

3626 - VINTERS, H., DAINTY, J., TYREE, M.T.: Cell wall elastic properties of *Chara corallina*. - Can. J. Bot. 55: 1933-1939, 1977.

*3627 - VITKOV, M.: Vodopotreblenie na fasul, otglezhdan na opodzolen chernozem v seve-roiztochna B"lgaria. [Water requirements of beans grown on podzolized chernozem soil in northeast Bulgaria.] - Rasteniev"dni Nauki 12: 100-104, 1975. [In Bulg, ab: R, E.]

*3628 - VOROBEĬKOV, G.A.: Vliyanie khlorkholinkhlorida na ustoĭchivost' pshenitsy k izbytochnomu i nedostatochnomu vodosnabzheniyu v kriticheskiĭ period. [Effect of CCC on resistance of wheat to excess and deficiency of water supply during critical period.] - Fiziol. Rast. 23: 573-578, 1976. [In R, ab: E.]

3629 - VOROBEĬKOV, G.A., ANIKINA, R.D.: Vliyanie regulyatorov rosta na ustoĭchivost' soi k pochvennomu zatopleniyu. [Effect of growth regulators on resistance of soya to soil flooding.] - Fiziol. Rast. 24: 1269-1275, 1977. [In R, ab: E.]

*3630 - VOROBEĬKOV, G.A., ANIKINA, R.D.: Obrazovanie kluben'kov i vlagoustoĭchivost' rasteniĭ soi pod vliyaniem obrabotki regulyatorami rosta. [Influence of growth regulators on formation of tubercles and resistance to waterlogging of soy-bean plants.] - In: Ustoĭchivost' Rasteniĭ k Pereuvlazhneniyu Pochvy v Uslo-viyakh Dal'nego Vostoka. Pp. 71-82. DVGU, Vladivostok 1976. [In R.]

3631 - VRKOČ, F.: K dynamice růstu a produktivitě hlavních polních plodin. [Contri-bution to growth dynamic and productivity of principal field crops.] - In: Produkce Biomasy a Tvorba Výnosu Polních Plodin. Vol. 1. Pp. 13-23. ČVTSZ, Praha 1977. [In Czech, ab: R, E, G.]

*3632 - WALCOTT, J.J., LAING, D.R.: Some physiological aspects of growth and yield in wheat crops: A comparison of a semidwarf and a standard height cultivar. - Aust. J. exp. Agr. anim. Husb. 16: 578-587, 1976.

3633 - WALTON, D.C., GALSON, E., HARRISON, M.A.: The relationship between stomatal resistance and abscisic-acid levels in leaves of water stressed bean plants. - Planta 133: 145-148, 1977.

3634 - WARRINGTON, I.J., PEET, M., PATTERSON, D.T., BUNCE, J., HASLEMORE, R.M., HELLMERS, H.: Growth and physiological responses of soybean under various thermoperiods. - Aust. J. Plant Physiol. 4: 371-380, 1977.

3635 - WATTS, W.R.: Field studies of stomatal conductance. - In: LANDSBERG, J.J., CUTTING, C.V. (ed.): Environmental Effects on Crop Physiology. Pp. 173-189. Academic Press, London - New York - San Francisco 1977.

3636 - WEBSTER, B.D., LEOPOLD, A.C.: The ultrastructure of dry and imbibed cotyledons of soybean. - Amer. J. Bot. 64: 1286-1293, 1977.

3637 - WEBSTER, R., HODGE, C.A.H., DRAYCOTT, A.P., DURRANT, M.J.: The effect of soil type and related factors on sugar beet yield. - J. agr. Sci. 88: 455-469, 1977.

3638 - WEISSHAUPT, F., REIIIER, W., NÜTZMANN, G.: Abhängigkeit des Pflanzenertrages von der Höhe der Zusatzwassergaben und deren flächenhafter Verteilung. - Arch. Acker- Pflanzenbau Bodenk. 21: 899-907, 1977.

3639 - WELLS, B.R., GILMOUR, J.T.: Sterility in rice cultivars as influenced by MSMA rate and water management. - Agron. J. 69: 451-454, 1977.

3640 - WENDT, C.W., ONKEN, A.B., WILKE, O.C., HARGROVE, R., BAUSCH, W., BARNES, L.: Effects of irrigation systems on the water requirements of sweet corn. - Soil Sci. Soc. Amer. J. 41: 785-788, 1977.

3641 - WIEBE, H.H., BROWN, R.W., BARKER, J.: Temperature gradient effects on in situ hygrometer measurements of water potential. - Agron. J. 69: 933-939, 1977.

3642 - WIEBE, H.H., PROSSER, R.J.: Influence of temperature gradients on leaf water potential. - Plant Physiol. 59: 256-258, 1977.

3643 - WIEBE, H.J., SCHÄTZLER, H.P., KÜHN, W.: On the movement and distribution of calcium in white cabbage in dependence of the water status. - Plant Soil 48: 409-416, 1977.

*3644 - WIEDENROTH, E.-M.: Chlorophyllbildung und Gaswechsel von Weizenkeimpflanzen (*Triticum aestivum* L.) unter dem Einfluss variierter Temperatur- und Sauerstoffbedingungen im Wurzelbereich. - Tagungsber. Akad. Landwirtschaftswiss. DDR, Berlin 143: 297-309, 1976.

3645 - WILCOX, G.E., MITCHELL, C.A., HOFF, J.E.: Influence of nitrogen form on exudation rate, and ammonium amide, and cation composition of xylem exudate in tomato. - J. Amer. Soc. hort. Sci. 102: 192-196, 1977.

3646 - WILKINSON, R.E., KARUNEN, P.: Water utilization by S-ethyl diprophylthiocarbamate treated wheat. - Weed Res. 17: 335-338, 1977.

3647 - WILL, G.M.: A field lysimeter to study water movement and nutrient content in a pumice soil under *Pinus radiata* forest. I. Site and construction details. - N. Zeal. J. Forest Sci. 7: 144-150, 1977.

*3648 - WILLIAMS, J.H., WILSON, J.H.H., BATE, G.C.: The growth of groundnuts (*Arachis hypogaea* L. cv. Makulu Red) at three altitudes in Rhodesia. - Rhodesian J. agr. Res. 13: 33-43, 1975.

*3649 - WILLIAMS, N.H.: Subsidiary-cell development in the *Catasetinae* (*Orchidaceae*) and related groups. - Bot. J. Linn. Soc. 72: 299-309, 1976.

3650 - WILLIAMS, W.T., BOUNDY, C.A.P., MILLINGTON, A.J.: The effect of sowing date on the growth and yield of three sorghum cultivars in the Ord River valley. II. The components of growth and yield. - Aust. J. agr. Res. 28: 381-387, 1977.

*3651 - WILLIS, W.O.: Soil water management and other factors that affect crop production. - In: Efficiency of Water and Fertilizer Use in Semi-Arid Regions. A Technical Document. Pp. 1-17. International Atomic Energy Agency, Vienna 1976.

3652 - WILLMER, C.M., RUTTER, J.C.: Guard cell malic acid metabolism during stomatal movements. - Nature 269: 327-328, 1977.

3653 - WILSON, J.R.: Variation of leaf characteristics with level of insertion on a grass tiller. III. Tissue water relations. - Aust. J. Plant Physiol. 4: 733-743, 1977.

*3654 - WINTER, K., TROUGHTON, J.H., EVENARI, M., LÄUCHLI, A., LÜTTGE, U.: Mineral ion composition and occurence of CAM-like diurnal malate fluctuations in plants of coastal and desert habitats of Israel and the Sinai. - Oecologia 25: 125-143, 1976.

3655 - WISBEY, B.D., BLACK, T.A., COPEMAN, R.J.: Controlling soil water matric potential in root disease studies. - Can. J. Bot. 55: 825-830, 1977.

3656 - WIT, C.T. de,: Modelle der Ertragsbildung als Brücke zwischen Prozess und System. - In: UNGER, K. (ed.): Biophysikalische Analyse Pflanzlicher Systeme. Pp. 19-30. VEB Gustav Fischer Verlag, Jena 1977.

3657 - WOLEDGE, J.: The effects of shading and cutting treatments on the photosynthetic rate of ryegrass leaves. - Ann. Bot. 41: 1279-1286, 1977.

3658 - WOLFF, J.O., WEST, S.D., VIERECK, L.A.: Xylem pressure potential in black spruce in interior Alaska. - Can. J. Forest Res. 7: 422-428, 1977.

3659 - WOLGAST, L.J., STOUT, B.B.: The effects of relative humidity at the time of flowering on fruit set in bear oak (*Quercus ilicifolia*). - Amer. J. Bot. 64: 159-160, 1977.

3660 - WONG, S.C., HEW, C.S.: Diffusive resistance, titratable acidity, and $CO_2$ fixation in two tropical epiphytic ferns. - Amer. Fern J. 66: 121-123, 1977.

3661 - WRENCH, P., WRIGHT, L., BRADY, C.J., HINDE, R.W.: The source of carbon for proline synthesis in osmotically stressed artichoke tuber slices. - Aust. J. Plant Physiol. 4: 703-711, 1977.

3662 - WRIGHT, S.T.C.: The relationship between leaf water potential ($\psi_{leaf}$) and the levels of abscisic acid and ethylene in excised wheat leaves. - Planta 134: 183-189, 1977.

3663 - WYATT, J.E.: Seed coat and water absorption properties of seed of near-isogenic snap bean lines differing in seed coat color. - J. Amer. Soc. hort. Sci. 102: 478-480, 1977.

3664 - WYN JONES, R.G., STOREY, R., LEIGH, R.A., AHMAD, N., POLLARD, A.: A hypothesis
on cytoplasmatic osmoregulation. - In: MARRÈ, E., CIFERRI, O. (ed.): Regulation
of Cell Membrane Activities in Plants. Pp. 121-136. Elsevier/North-Holland
Biomedical Press, Amsterdam - Oxford - New York 1977.

*3665 - YABUKI, K., KIYOTA, M.: Studies on the carbon dioxide environment for plant
growth. VI. Effects of the carbon dioxide concentration on transpiration rate.
- Environ. Control Biol. 13: 151-158, 1975.

*3666 - YELENOSKY, G.: Cold hardening in citrus stems. - Plant Physiol. 56: 540-543,
1975.

3667 - YOUNG, R.H., GARNSEY, S.M.: Water uptake patterns in blighted citrus trees. -
J. Amer. Soc. hort. Sci. 102: 751-756, 1977.

3668 - ZAHRAN, M.A., EL-BAGOURY, I.H., ABDEL-WAHID, A.A., EL-DEMERDASH, M.A.: Trans-
plantation of *Juncus* spp. on saline soils in Egypt. - In: DREGNE, H.E. (ed.):
Managing Saline Water for Irrigation. Pp. 142-154. Texas Technical University,
Lubbock 1977.

3669 - ZANSTRA, P.E., HAGENZIEKER, F.: Comments on the psychrometric determination of
leaf water potentials in situ. - Plant Soil 48: 347-367, 1977.

3670 - ZAROGIANNIS, V.: Der Einfluss der Krumenmächtigkeit des Bodeus auf den Ertrag
und die Qualität der Zuckerrübe bei vier Beregnungsvarianten. - Bodenkultur
28: 386-404, 1977.

3671 - ZEEVAART, J.A.D.: Sites of abscisic acid synthesis and metabolism in *Ricinus
communis* L. - Plant Physiol. 59: 788-791, 1977.

3672 - ZEIGER, E., HEPLER, P.K.: Light and stomatal function: Blue light stimulates
swelling of guard cell protoplasts. - Science 196: 887-889, 1977.

3673 - ZEIGER, E., MOODY, W., HEPLER, P., VARELA, F.: Light-sensitive membrane potentials in onion guard cells. - Nature 270: 270-271, 1977.

3674 - ZEMÁNEK, M.: Spolupůsobení vody a dusíku při tvorbě výnosu jarního ječmene.
[Interaction of water and nitrogen in formation of spring barley crop.] -
Rost. Výroba 23: 683-693, 1977. [In Czech, ab: R, E, G.]

*3675 - ZHURAVLEVA, N.A.: Ekologo-fiziologicheskaya kharakteristika stepnykh rasteriĭ
i ikh zasukhoustoĭchivost'. [Ecophysiological characteristics of steppe plants
and their resistance to drought.] - In: Vodnyĭ Obmen v Osnovnykh Tipakh Rasti-
tel'nosti SSSR kak Element Krugovorota Veshchestva i Énergii. Pp. 160-165. Nauka, Novosibirsk 1975. [In R.]

3676 - ŽILA, L.: Niektoré fyziologické procesy cukrovej repy pri róznych metódach
závlahového režimu. [Some physiological processes of sugar beet irrigated according to different methods.] - Rost. Výroba 23: 695-703, 1977. [In Slov,
ab: E, G, R.]

3677 - ŽILA, L., HUZULÁK, J., LEDEČOVÁ, B.: Metodický príspevok k moraniu charakteris-
tík vodného režimu marhule a broskyne. [Methodical contribution to measurement
of water regime characteristics in apricot and peach.] - In: HUZULÁK, J., MASA-
ROVICOVÁ, E. (ed.): Fotosyntéza a Vodný Režim Drevín. Pp. 115-122. Modra-Piesky 1977. [In Slov, ab: E, R.]

3678 - ZIMA, M.: Vzťah medzi obsahom vody a fotosyntézou listov topoľa. [The relationship between water content and photosynthesis of poplar leaves (*Populus euro-
americana*).] - In: HUZULÁK, J., MASAROVICOVÁ, E. (ed.): Fotosyntéza a Vodný
Režim Drevín. Pp. 99-105. Modra-Piesky 1977. [In Slov, ab: E, R.]

3679 - ZIMA, M.: Možnosti regulácie fyziologickej aktivity odnoží obilnin. [Possibil-
ity of physiological activity regulation of grain crop tillers.] - In: REPKA,
J. (ed.): Zborník Referátov zo Seminára Fyziologicko-Genetické a Chemické Faktory Produktivity Rastlín. Pp. 152-159. VŠP, Nitra 1977. [In Slov.]

3680 - ZIMMERMANN, U.: Cell turgor pressure regulation and turgor pressure-mediated
transport processes. - In: JENNINGS, D.H. (ed.): Integration of Activity in the
Higher Plant. Pp. 117-154. Cambridge University Press, Cambridge - London - New
York - Melbourne 1977.

3681 - ZIMMERMANN, U., BECKERS, F., COSTER, H.G.L.: The effect of pressure on the
electrical breakdown in the membranes of *Valonia utricularis*. - Biochim. biophys. Acta 464: 399-416, 1977.

3682 - ZLOTNIKOVA, I.F., GUNAR, I.I., PANICHIN, L.A.: Izmerenie vnutrikletochnoĭ
aktivnosti kaliya v kletkakh epidermisa lista tradeskantsii. [Measuring intracellular activity of potassium in epidermal cells of *Tradescantia albiflora*.]
- Izv. TSKHA 2: 10-16, 1977. [In R, ab: E.]

3683 - ZUBKOVA, I.G.: Modusy obrazovaniya ust'ichnogo apparata lista v sem. *Ranunculaceae* Juss. [Moduses of formation of leaf stomata in the family *Ranunculaceae*
Juss.] - Bot. Zh. 62: 1179-1182, 1977. [In R.]

3684 - ZURZYCKI, J., METZNER, H.: Volume changes of chloroplasts *in vivo* at high den-
sities of blue and red radiation. - Photosynthetica 11: 260-267, 1977.

3685 - ZYALALOV, A.A.: O roli épidermisa v regulyatsii vodoobmena. [The role of epidermis on regulation of water exchange.] - Fiziol. Rast. 24: 779-784, 1977.
[In R, ab: E.]

3686 - ZYALALOV, A.A.: O pokazatelyakh  sostoyaniya vody i opredelenii depressii ee
khimicheskogo potentsiala. [Parameters of the state of water and determination
of depression of water chemical potentials.] - Fiziol. Rast. 24: 1247-1250,
1977. [In R, ab: E.]

## AUTHORS' INDEX

Authors' names are presented in the form in which they appear in the respective publication. The names from papers published in Cyrillic character are transcribed as
shown in Instructions for Use. Alternative spelling and form of the name of the same
author are usually cross-indexed.

**A**

ABAZA, M.   2940
ABDEL-WAHID, A.A.   3668
ABROL, I.P.   2813
ACEVES-N, E.   2480
ACHARYYA, N.   3586
ACKERSON, R.C.   2481, 2482, 2483,
2484
ADAMS, J.A.   2485
ADAMSON, R.M.   2517
ADRIANO, D.C.   3295
AFSCHAR, I.   2818
AGARWAL, M.C.   3609
AGARWAL, S.K.   2486
AGATA, W.   3552
AHARONI, N.   2487
AHLGRIMM, H.-J.   2488
AHMAD, I.   2489
AHMAD, N.   3664
AHO, N.   2490
AKITA, S.   3550
AKSENOV, S.I.   2491, 2492, 3201
AKSYONOV, S.I.   2493
ALBERTE, R.S.   2494, 2606, 3245
ALBRIGO, L.G.   2495
ALEKSEEVA, V.Ya.   2496
ALESSI, J.   2497, 2498
ALFANI, A.   3074
ALFARO, J.F.   2871
ALI, H.C.   2499
ALJIBURY, F.K.   3247
ALLEN, J.F.   2500
ALLEN, L.H.,Jr.   2501
ALLEN, R.R.   3181
ALMADI, L.   2502
AL-SAADI, H.A.   2503
ALVAREZ, E.I.   2504
ALVAREZ, J.M.   2732
AMIRDZHANOV, A.G.   2505
ANDO, T.   3532
ANDREWS, P.   2506
ANGUS, J.F.   2507
ANIKINA, R.D.   3629, 3630
ANTIPOV, N.I.   2508, 2509
APPUKUTTAN, E.   2553
ARDAKANI, M.S.   2510, 2788
ARDEKANI, E.R.   3420
ARDITTI, J.   2836
ARIHARA, J.   2511

ARKHANGEL'SKAYA, M.A.   3044
ARMSTRONG, R.A.   2512
ARNOLD, W.N.   2513
ARUNAKUMARI, P.   3422
ASBELL, C.W.   2877
ASHLEY, D.A.   2551
ASHUROV, A.A.   2514
ASKOCHENSKAYA, N.A.   2492
ASLAM, M.   2515
ASPINALL, D.   3515
ASTHANA, D.C.   2516
ASTON, M.J.   2858
ATKINS, C.A.   2782, 3241
ATKINSON, R.G.   2517
AUCLAIR, D.   2518
AVADHANI, P.N.   2836
AVRIGEANU, G.   3466
AYERBE, L.   2519
AYERS, R.S.   2520, 3194
AYRES, P.G.   2521, 2522

**B**

BAGGA, A.K.   3335
BAGLEY, J.O.   2899
BAÏDULOVA BABKO, T.Yu   3531
BAIJAL, B.D.   3448, 3611, 3612, 3613
BAINES, M.A.   2678
BAINS, S.S.   3037
BAJPAI, M.R.   2523
BAJWA, M.S.   3305
BAKER, C.J.   2524
BALASUBRAMANIAN, V.   3320
BALL, E.   3078, 3079, 3080
BALLARD, L.A.T.   2525
BALLAUX, J.C.   3163
BALLIO, A.   2526
BANGLIANG, S.   3534
BAÑOCH, Z.   2527
BARA, M.   2528
BARADAS, M.W.   2529
BARTHOLIC, J.F.   3361
BARKER, J.   3641
BARKHAM, J.P.   2530
BARLOW, E.W.R.   2531, 2532
BARNES, L.   3640
BARNETT, A.P.   2624

PETINOV, N.S.   3270
PETKOV, P.S.   3271, 3272
PETRASOVITS, I.   3273
PETRENKO, N.I.   3571
PETROV, A.P.   3274
PETROVA, L.I.   2993
PETROVA, V.N.   3275
PEVELING, E.   3276
PFENDER, W.F.   3277
PHAM THI, A.T.   3278
PHENE, C.J.   2620
PHILLIPS, R.D.   3279
PHIPPS, P.M.   3280
PIERI, C.   3281
PILL, W.G.   3282
PITMAN, M.G.   3283, 3284
PITT, J.I.   3285
PLAMENAC, N.   3286
PLATT-ALOIA, K.   2877
PLAUT, Z.   3287
POKORNÝ, V.   2824
POLERECKÝ, O.   3288
POLLARD, A.   3664
POLLARD, D.F.W.   3289
POLLMER, W.G.   2800
POPOV, B.A.   3007
PORTAS, C.A.M.   3290
PORTER, N.G.   3040
POSPISHILOVA, Ya.   3291 see also
POSPÍŠILOVÁ, J.   3292, 3293, 3488
POWELL, D.B.B.   3294
POWER, J.F.   2497, 2498
POWERS, W.L.   2966, 3100
PRATT, P.F.   3295
PRAŽAK, M.   3296
PREMECZ, G.   3297
PRESOLY, E.   3123
PRESSLAND, A.J.   3298
PRESSMAN, E.   3299
PRIEHRADNÝ, S.   3300, 3301, 3302, 3303
PRIETO, J.M.J.   2925
PRIHAR, S.S.   3304, 3305, 3390
PRIOUL, J.-L.   2657
PROEBSTING, E.L.   3306
PROKHORCHIK, R.A.   3307
PRONINA, N.D.   2846
PROSSER, R.J.   3642
PRUITT, W.O.   3517
PTÁČKOVÁ, M.   3308, 3309, 3310
PUJOL, B.   3311
PUNTAMKAR, S.S.   3423
PURI, J.   3312


## Q

QUARRIE, S.A.   3313
QUISENBERRY, J.E.   3350, 3351


## R

RAATS, P.A.C.   3314
RACUSEN, R.H.   3315
RADOMSKI, C.   3316
RAGHAVENDRA, A.S.   2690, 3317, 3332
RAGHU, J.S.   3318, 3427
RAGUSE, C.A.   3319
RAI, R.   3190
RAINS, D.W.   2680, 2681, 2682
RAINY, M.E.   2530
RAJAGOPAL, V.   3320
RAJENDRA, B.R.   3321
RAJESWARI, V.M.   2868
RAJPUT, R.K.   2930
RAJU, V.S.   3322, 3324
RAMACHANDRAM, M.   3326
RAMAKRISHNA, Y.S.   3463
RAMAYYA, N.   3325
RAND, R.H.   2639, 2667, 3323
RANDAZZO, G.   2526
RANEY, R.J.   3365
RANI, S.   2868
RAO, I.M.   2690
RAO, M.S.R.M.   3326
RAO, P.N.   3322, 3324
RAO, S.R.S.   3325
RAO, T.V.   3097
RAO, V.R.   3326
RASCHKE, K.   2713, 2714, 3327, 3328, 3329
RASMUSSEN, V.P.   2899
RATHAIAH, Y.   3330, 3331
RATHNAM, C.K.M.   3332
RATNAM, B.P.   3333
RATNASOORIYA, G.B.   3182
RAUSCHKOLB, R.S.   3247
RAUZI, F.   2886
RAVEN, J.A.   3334
RAVINDRANATH, E.   3235
RAWSON, H.M.   3335, 3336
RAY, S.   3037
READ, D.J.   2811
REBELLA, C.   3560
REDMANN, R.E.   3349
REESE, R.L.   3004, 3337, 3338
REGINATO, R.J.   2938, 2950, 3339
REHM, G.W.   3340
REICOSKY, D.C.   3341, 3342
REID, C.P.P.   3343
REIHER, W.   3638
REISNER, A.H.   2532
RHOADES, J.D.   3217
RICHARDS, D.   3344
RICHMOND, A.E.   2487, 2580
RICHTER, H.   3029, 3345
RICKARD, W.H.   2660
RIEKELS, J.W.   3346
RIES, S.K.   3347
RIJKS, D.A.   3348
RIKIN, A.   2580
RIPLEY, E.A.   3349

# PLANT INDEX

This index contains plant genera and types interesting as experimental material for physiological, ecological and agricultural studies. The Latin plant names are the main items which present the reference number. English names of the most common plants are cross-indexed.

carrot see *Daucus*

*Carthamus*  2536, 2728, 2998, 3183, 3333, 3419

*Carya*  2761

*Cassia*  3174

*Cattleya*  2836

*Ceanothus*  3012, 3064

cedar see *Tamarix*

*Cedrus*  3599

*Cerasus*  2540

*Chenopodium*  3539

cherry see *Cerasus*

*Chloris*  3611

*Chrysanthemum*  2817, 3399

*Chrysosplenium*  3164

*Cicer*  2872

*Cistus*  3113

*Citrullus*  2552, 3437

*Citrus*  2495, 2533, 2589, 2981, 3004, 3061, 3086, 3107, 3108, 3109, 3110, 3337, 3338, 3360, 3361, 3437, 3551, 3572, 3666, 3668

*Clethra*  3403

clover see *Trifolium*

*Cocos*  2533, 3144, 3145

*Codiaeum*  3324

*Coffea*  2533, 2598, 2611, 3103, 3484, 3557

*Colchicum*  3379

*Coleus*  3369

*Commelina*  2713, 2714, 3329, 3512, 3567, 3652

*Comptonia*  2761

*Convallaria*  2734

cornel see *Cornus*

*Cornus*  2642, 2643, 2698, 2734, 2761, 2816, 3262, 3356

*Cotoneaster*  2514

cotton see *Gossypium*

cowpea see *Vigna*

cranberry see *Vaccinium*

*Crataegus*  2734

*Crithmum*  3654

*Crotalaria*  2941, 3083

*Croton*  2691, 3324

*Cryptomeria*  3058

cucumber see *Cucumis*

*Cucumis*  2545, 2636, 2661, 2718, 2770, 2773, 2809, 3367, 3383, 3561, 3665

*Cucurbita*  2880

*Curatella*  3393

*Cymbopogon*  2544

*Cynodon*  3195


**D**

*Dactylis*  2596, 2838, 3416

*Danthonia*  2876

*Datura*  3044, 3315, 3317

*Daucus*  2897, 3334, 3357, 3367, 3470, 3655

*Dendrophthoe*  3024

*Deschampsia*  2633, 3005

*Desmodium*  3284, 3441

*Dianthus*  2751

*Dieffenbachia*  3399

*Digera*  2691

*Digitalis*  3044

*Digitaria*  2767, 3552

dogwood see *Cornus*

Douglas fir see *Pseudotsuga*

# SUBJECT INDEX

This index contains a selection of primary items chosen according to their interest for water relation researchers and to their relative importance and occurence.


## A

Bound water  2491, 2492, 2643, 2692, 2989, 2992, 2996, 3028, 3188, 3201, 3231, 3257, 3356, 3407, 3675

Bound water, methods  2491

Boundary layer see Conductance for water vapour and $CO_2$ transfer, boundary layer


C

Canopy architecture see Flooding, ...; Irrigation, ...; Precipitation and dew, effect on canopy architecture

Canopy model see Model of canopy

Carbohydrates, relation to stomata and epidermis  2817,  3444

Carbohydrates, relation to water status in plant  3148

Carbohydrates, relation to wilting  2538, 2652, 3619

Carbon fixation pathways see Irrigation, ...; Osmotic agents, ...; Salinity, ...; Soil moisture, ...; Water status in plant, effect on carbon fixation pathways

Carbowax see Osmotic agents,...

Carotenoids see Soil moisture, effect on carotenoids

Chlorophyll see Drought, ...; Flooding, ...; Humidity of air, ...; Irrigation, ...; Osmotic agents, ...; Precipitation and dew, ...; Salinity, ...; Soil moisture, ...; Water status in plant, effect on chlorophyll

Chloroplasts see Drought, ...; Humidity of air, ...; Osmotic agents, ...; Soil moisture, ...; Water status in plant, effect on chloroplasts

$CO_2$, effect on conductance for water vapour and $CO_2$ transfer  2658, 2751, 2773, 2861, 2863, 3020, 3443, 3568

$CO_2$, effect on stomata and epidermis  2673, 2677, 2731, 2801, 2942, 3170, 3255, 3327, 3439, 3443, 3475, 3575

$CO_2$, effect on transpiration  2673, 2975, 3020, 3093, 3665

$CO_2$, effect on water absorption  3141, 3665

$CO_2$ influx see Drought, ...; Flooding, ...; Humidity of air, ...; Irrigation, ...; Osmotic agents, ...; Precipitation and dew, ...; Salinity, ...; Soil Moisture, ...; Water status in plant, effect on $CO_2$ influx

Conductance for $CO_2$ transfer, epidermis  2485, 2551, 2614, 2631, 2663, 2677, 2751, 2973, 3068, 3072, 3206, 3245, 3252, 3323, 3556

Conductance for $CO_2$ transfer, mesophyll (intracellular)  2485, 2515, 2531, 2541, 2545, 2551, 2565, 2603, 2626, 2663, 2719, 2766, 2833, 2835, 2923, 2941, 2973, 2986, 3020, 3021, 3067, 3068, 3106, 3156, 3157, 3206, 3219, 3256, 3336, 3392, 3396, 3405, 3474, 3475, 3568, 3571, 3597, 3657

Conductance for water vapour and $CO_2$ transfer see also Age of plant, ...; $CO_2$ ...; Cultivars, ...; Defoliation, decapitation, ...; Drought, ...; Farming practices, ...; Flooding, ...; Growth substances, hormones, inhibitors etc., ...; Humidity of air, ...; Irradiance, ...; Irrigation, ...; Leaf insertion level, ...; Mineral elements. ...; Osmotic agents, ...; Oxygen, ...; Pathological effect, ...; Pollutants and ozone, ...; Salinity, ...; Soil moisture, ...; Taxons, ...; Temperature, ...; Water status in plant, ...; Wind, effect on conductance for water vapour and $CO_2$ transfer

Conductance for water vapour and $CO_2$ transfer, above canopy  2843, 2977

Conductance for water vapour and $CO_2$ transfer, boundary layer  2658, 2677, 2833, 2843, 3041, 3067, 3068, 3106, 3218, 3392, 3449, 3475, 3568, 3597, 3620, 3624

Conductance for water vapour and $CO_2$ transfer, canopy  2631, 2842, 2977, 3005, 3067, 3349, 3397, 3449, 3450, 3472, 3555

Conductance for water vapour and $CO_2$ transfer, comparison of plants with different types of carbon metabolism  2691

# I

Model  2599, 2619, 2636, 2639, 2646, 2648, 2649, 2654, 2658, 2667, 2677, 2679, 2687,
    2705, 2760, 2764, 2772, 2779, 2780, 2788, 2796, 2810, 2812, 2838, 2871, 2883,
    2899, 2901, 2903, 2908, 2909, 2910, 2919, 2924, 2964, 2977, 2978, 3041, 3042,
    3062, 3067, 3081, 3122, 3127, 3178, 3183, 3210, 3217, 3248, 3252, 3254, 3258,
    3260, 3266, 3270, 3286, 3294, 3323, 3349, 3361, 3365, 3378, 3421, 3431, 3468,
    3471, 3473, 3474, 3500, 3501, 3555, 3563, 3588, 3616, 3620, 3635, 3656, 3664

Model of canopy  2918, 3469, 3472, 3581, 3590


# N

Nitrogen, effect on stomata and epidermis  2575


# O

$O_2$ see Oxygen, ...

$O_3$ see Pollutants and ozone, ...

Ontogenesis see Age of plant, ...

Oscillations see Conductance for water vapour and $CO_2$ transfer, ...;. Stomatal aper-
    ture, ...; Transpiration rate, ...; Water status in plant, oscillations

Osmotic agents, effect on carbon fixation pathways  2568, 2640, 3056

Osmotic agents, effect on chlorophyll  2972, 3056

Osmotic agents, effect on chloroplasts  2809, 3153, 3237, 3684

Osmotic agents, effect on $CO_2$ influx  2500, 2640, 2952, 3278, 3401

Osmotic agents, effect on conductance for water vapour and $CO_2$ transfer  3633

Osmotic agents, effect on electron transport chain  2548, 2848, 3278

Osmotic agents, effect on growth and productivity  2489, 2536, 2556, 2571, 2581,
    2830, 2872, 2876, 2885, 3039, 3171, 3285, 3372, 3506

Osmotic agents, effect on leaf anatomy  2885

Osmotic agents, effect on photorespiration  3278

Osmotic agents, effect on respiration  3401

Osmotic agents, effect on transpiration  2522, 3067

Osmotic agents, effect on water absorption  2872, 3055, 3141, 3343

Osmotic agents, effect on water status in plant  2522, 2885, 3141, 3292, 3391, 3436,
    3570, 3633

Osmotic agents, effect on water transport in cells  3232, 3233, 3626

Osmotic agents, effect on wilting  2885, 3633

Osmotic potential in plant tissue (see also Water status in plant, ...)  2483, 2484,
    2511, 2522, 2541, 2545, 2546, 2552, 2576, 2919, 2626, 2627, 2628, 2665, 2676,
    2681, 2682, 2725, 2786, 2844, 2891, 2910, 2926, 2954, 3029, 3079, 3103, 3104,
    3113, 3145, 3148, 3162, 3166, 3167, 3175, 3222, 3228, 3229, 3231, 3284, 3292,
    3293, 3294, 3328, 3356, 3391, 3417, 3464, 3465, 3486, 3488, 3562, 3653, 3663,
    3664, 3677, 3680

Osmotic potential in substrate  2480, 2485, 2629, 2830, 2832, 2855, 2869, 2910, 2952,
    2964, 2965, 2993, 3091, 3176, 3417, 3527

Osmotic potential, methods  3013, 3029, 3592

Oxygen, effect on water vapour and $CO_2$ transfer  3487

Oxygen, effect on stomata and epidermis  2673, 3487

Oxygen, effect on transpiration  2673, 2767, 3625

Oxygen, effect on water absorption by plant  2480

Structure of cuticle  2574, 2783, 2785, 2982, 2999, 3012, 3083, 3554

Structure of epidermis 2604, 2635, 2681, 2712, 2773, 2783, 2784, 2814, 2999, 3012,
    3054, 3083, 3187, 3322, 3324, 3329, 3384, 3422, 3621

Surface impression technique see Stomatal aperture, methods, microrelief methods


T

Taxons, comparison of conductance for water vapour and $CO_2$ transfer  2605, 3006, 3262,
    3263, 3433, 3480

Taxons, comparison of stomata and epidermis  3263

Taxons, comparison of transpiration  2509, 2545, 2605, 2789, 2960, 2961, 3018, 3092,
    3480

Taxons, comparison of water status in plant  2509, 2545, 2605, 3113, 3263, 3356, 3403,
    3433, 3564, 3654

Temperature, effect on conductance for water vapour and $CO_2$ transfer  2545, 2735,
    2862, 2941, 2986, 3020, 3053, 3061, 3160, 3250, 3261, 3262, 3634, 3635

Temperature, effect on stomata and epidermis  2611, 2677, 2687, 2801, 2862, 3070,
    3250, 3255, 3411, 3438, 3475, 3550, 3634, 3635

Temperature, effect on transpiration  2531, 2545, 2578, 2789, 2833, 2890, 2933, 2941,
    2950, 2975, 2979, 2986, 2989, 3020, 3041, 3093, 3205, 3250, 3273, 3393, 3438,
    3536, 3550, 3620

Temperature, effect on water absorption by plant  2669, 2980, 3139

Temperature, effect on water status in plant  2503, 2531, 2562, 2595, 2729, 2900,
    2933, 2979, 2989, 3061, 3229, 3231, 3261, 3262, 3416, 3487, 3525, 3526, 3634,
    3642, 3677

Temperature, effect on water transport in cells 3228, 3229, 3230

Temperature, effect on water transport in plant  2979, 3525

Temperature, effect on wilting  2549, 3296, 3399, 3551, 3592

Transpiration see Age of plant, ...; $CO_2$, ...; Cultivars, ...; Drought, ...; Ecotypes,
    ...; Enzyme inhibitors, ...; Farming practices, ...; Flooding, ...; Genetics, ...;
    Growth substances, hormones, inhibitors etc., ...; Humidity of air, ...; Irradi-
    ance, ...; Irrigation, ...; Leaf insertion level, ...; Mineral elements, ...;
    Osmotic agents, ...; Oxygen, ...; Pathological effect, ...; Pesticides and
    herbicides, ...; Pollutants and ozone, ...; Precipitation and dew, ...; Salinity,
    ...; Soil moisture, ...; Taxons, ...; Temperature, ...; Water status in plant,
    ...; Wind, effect on transpiration

Transpiration chambers  3013, 3093, 3146, 3546

Transpiration coefficient  2497, 2505, 2582, 2616, 2647, 2691, 2702, 2789, 2813, 2838,
    2916, 2930, 2968, 3011, 3037, 3091, 3161, 3177, 3234, 3256, 3335, 3373, 3414,
    3459, 3461, 3463, 3477, 3530, 3586, 3596

Transpiration curves  2657, 2942, 2951, 3606

Transpiration, cuticular  2877, 2931

Transpiration, effect of antitranspirants see Antitranspirants

Transpiration integrated  2579, 2582, 2618, 2683, 2799, 2895, 2913, 2956, 2975, 2988,
    3002, 3018, 3062, 3072, 3092, 3156, 3157, 3204, 3256, 3287, 3332, 3365, 3537,
    3591

Transpiration rate and stomata see Stomata and transpiration rate

Transpiration rate, comparison of plants with different types of carbon metabolism
    2691, 2789

Transpiration rate, diurnal course  2490, 2545, 2550, 2578, 2636, 2677, 2729, 2802,
    2812, 2884, 2951, 2979, 3005, 3017, 3041, 3048, 3061, 3074, 3093, 3127, 3159,

Wilting (cont.) Oxygen, ...; Pathological effect,  ...; Pesticides and herbicides,
     ...; pH, ...; Precipitation and dew, ...; Proteins, amino acids, nucleic acids,
     ...; Salinity, ...; Soil moisture, ...; Temperature, ...; Wind, effect on wilting

Wilting, diurnal course  2550, 3351, 3417, 3440

Wilting, effect on ion absorption and transport  2538, 2629, 2711, 3033, 3060

Wilting, effect on other processes than above and below  2487, 2507, 2567, 2576, 2579,
     2600, 2621, 2642, 2643, 2682, 2684, 2694, 2695, 2696, 2709, 2767, 2768, 2827,
     2866, 2958, 3073, 3131, 3138, 3143, 3153, 3168, 3282, 3297, 3320, 3362, 3407,
     3430, 3451, 3494, 3510, 3515, 3579, 3662

Wilting, effect on plant metabolism  2482, 2494, 2538, 2564, 2583, 2645, 2704, 2710,
     2730, 2820, 2867, 2927, 2928, 2993, 3023, 3050, 3096, 3130, 3212, 3275, 3291,
     3312, 3358, 3408, 3409, 3514, 3661, 3671

Wilting, indicators, methods  2537, 2638, 2914, 3100, 3273, 3476, 3516, 3592, 3604,
     3605, 3640

Wilting, mechanisms of development, indicators  2482, 2487, 2494, 2501, 2531, 2538,
     2545, 2546, 2549, 2567, 2583, 2595, 2652, 2655, 2679, 2684, 2695, 2710, 2787,
     2801, 2804, 2806, 2807, 2831, 2892, 2906, 2923, 2933, 2954, 2957, 2977, 2979,
     2990, 3011, 3015, 3041, 3075, 3103, 3104, 3107, 3130, 3131, 3132, 3142, 3153,
     3167, 3231, 3276, 3278, 3292, 3293, 3388, 3417, 3451, 3464, 3465, 3505, 3522,
     3642, 3662, 3671

Wilting, seasonal course  2598, 2933, 3296, 3551, 3675

Wind, effect on conductance for water vapour and $CO_2$ transfer  2632, 2795, 2842, 2843,
     3405

Wind, effect on stomata and epidermis  2632, 2843, 2844

Wind, effect on transpiration  2632, 2843, 2886, 2933, 2951, 3005, 3620, 3665

Wind, effect on water absorption  2733, 3055, 3665

Wind, effect on water status in plant  2795, 2844, 2933

Wind, effect on wilting  2632


X

Xylem transport of water see Water transport in plant, transport in xylem ...